New Vanguard • 2

# M1 Abrams Main Battle Tank 1982–92

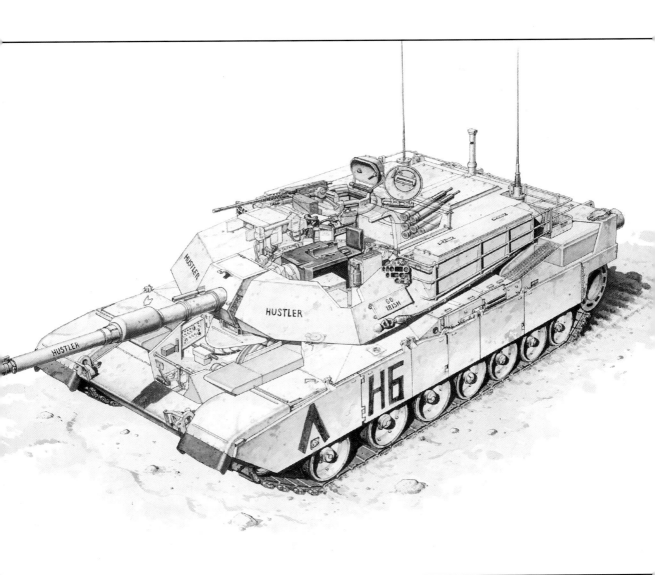

Steven J Zaloga • Illustrated by Peter Sarson

First published in Great Britain in 1993 by
Osprey Publishing, Midland House,
West Way, Botley, Oxford OX2 0PH, UK
443 Park Avenue South, New York, NY 10016, USA
Email: info@ospreypublishing.com

Reprinted 1995, 1997, 1998, 1999, 2001, 2002, 2003 (twice), 2004, 2005

All rights reserved. Apart from any fair dealing for the purpose of private study, research, criticism or review, as permitted under the Copyright Designs and Patents Act, 1988, no part of this publication may be reproduced, stored in a retrieval system, or transmitted in any form or by any means, electronic, electrical, chemical, mechanical, optical, photocopying, recording or otherwise, without the prior permission of the copyright owner. Enquiries should be addressed to the Publishers.

CIP Data for this publication is available from the British Library

ISBN 185532 283 8

Filmset in Great Britain
Printed in China through World Print Ltd.

FOR A CATALOGUE OF ALL BOOKS PUBLISHED BY
OSPREY MILITARY AND AVIATION PLEASE CONTACT:

NORTH AMERICA
Osprey Direct, 2427 Bond Street,
University Park, IL 60466, USA
E-mail: info@ospreydirectusa.com

ALL OTHER REGIONS
Osprey Direct UK, P.O. Box 140,
Wellingborough, Northants, NN8 2FA, UK
E-mail: info@ospreydirect.co.uk

www.ospreypublishing.com

## Acknowledgements

The author would like to thank a number of people for their help on this book Thanks to Stephen 'Cookie' Sewell, Pierre Touzin, Michael Jerchel, Joel Paskauskas, Bob Lessels, and Frank DeSisto for help with photographs and other material; to Lt. Col. John Craddock, and Capt. David Hubner of the 24th Infantry Division (M) for their time and patience in relating experiences on the M1A1 during Desert Storm; to Capt. Steve Hart, and Bill Rosenmund of US Army Public Affairs for helping obtain photographs; and also to General Dynamics Land Systems Division, BMY Corp. and Alliant Techsystems for photographs.

# M1 ABRAMS MAIN BATTLE TANK 1982-92

## DEVELOPMENTAL HISTORY

The M1 Abrams tank is the most radical departure in US tank design since World War 2. Until the advent of the M1 tank in the early 1980s, the US Army had relied on the steady evolution of the M26 Pershing tank: M46, M47, M48 and M60. Earlier attempts to replace this evolutionary cycle with a new design, such as the T-95 and MBT-70, had been failures. The M1 design came at a time when there were a host of important new tank technologies coming to fruition: special armours, thermal imaging sights, advanced gun fire controls, and turbine engines. These many new technologies were integrated into the M1 Abrams tank design. Unlike the MBT-70, the M1 was not designed to be the best tank in the world: it was designed to be the best tank possible within a limited budget. The M1A1 Abrams is a less expensive tank than its two nearest counterparts, the German Leopard II and British Challenger, but certainly a strong contender for the title of the world's best main battle tank.

The M1 Abrams was the culmination of US Army programmes begun in the 1960s to replace the M60 series of main battle tanks. The first major attempt of the programme was a joint German-American programme for a common main battle tank (MBT) which was called the MBT-70 by the US Army. The programme proved too sophisticated and too expensive, and was cancelled following Germany's withdrawal. A simplified version, designated XM803, was examined but it too was cancelled at the end of 1971. In February 1972, the MBT task force was established at Ft Knox under Maj.Gen. William Desobry to examine the requirements for a new tank. Contracts were awarded to Chrysler and General Motors to begin advanced design work on the XM815, as the new tank was initially designated.

The requirement objectives for the new tank had the following order of priorities: crew survivability; surveillance and target acquisition performance; first-round and subsequent hit probability; minimal time to acquire and hit; cross-country mobility; complementary armament integration; equipment survivability; crew environment; silhouette; acceleration and deceleration; ammunition stowage; human factors; productibility; range; speed; diagnostic maintenance aids; growth potential; support equipment; transportability.

Consideration was given to three weapons for the tank's main gun: the existing M68 105 mm gun, a British 110 mm rifled gun and a German Rheinmetall 120 mm smoothbore. It was decided to stay with the existing 105 mm gun due to several important new advances in projectile design. The new APFSDS[1] projectiles were significantly superior to existing ammunition in armour penetration and had the added advantage of maintaining standardisation with existing US Army and NATO tanks. The British 110 mm gun was judged insufficiently superior to the 105 mm gun, while the German gun was not expected to be ready in time to meet the production requirements

*The last gasp of the failed MBT-70 programme was the XM803. This tank was intended to be a low-cost, austere version of the overly sophisticated MBT-70. It too was cancelled in favour of a new tank. (Author)*

[1] Armour-piercing, fin-stabilised, discarding-sabot: a type of kinetic energy penetrator, more popularly called 'sabot' ammunition by US tank crews.

*The XM1 development programme was a competitive effort between General Motors and Chrysler. Seen here is the General Motors validation phase tank. (US Army)*

for the new XM815. The US Army was considering the development of a follow-on missile to the 152 mm Shillelagh, nicknamed 'Swifty' which could be fired from the 105 mm gun tube.

Several armament innovations were considered for the new tank. One option was to incorporate a 25 mm Bushmaster cannon in place of the co-axial machine gun. This was considered since tanks frequently encounter targets, such as light armoured vehicles or infantry inside buildings, which are too well protected to defeat with .50cal. machine gun fire. While such targets can be defeated by main gun fire, extensive use of the main gun against such targets would soon exhaust the tank's ammunition supply. A 25 mm autocannon could defeat such targets and help conserve main gun ammunition.

Another concern was to increase the survivability of the design to enemy tank fire. Historically, the primary cause of tank loss in diesel-powered tanks has been ammunition fires. To get around this problem, proposals were examined which would place the ammunition in a compartment at the rear of the turret to contain a fire in the event of the tank's armour being penetrated.

The US Army decided to take an unusual step in the developmental process by opting for competitive prototypes from Chrysler and General Motors. It was anticipated that competition would lead to better designs at a lower cost, and alternative design options could be examined as well. The General Motors design incorporated a diesel engine while the Chrysler prototype was fitted with an unconventional turbine engine. A major element of the new MBT programme was to reduce the unit cost of the tank compared to the failed MBT-70 programme and this requirement shaped which technologies could and could not be incorporated into the new tank.

Two events in 1973 were to have profound

*A rear view of the General Motors XM1. This vehicle had far less prominent exhaust venting at the rear than the Chrysler XM1 since it used the AVCR-1360 diesel engine. (US Army)*

*The Chrysler XM1 validation phase tank is seen here during the trials. This was the second turret configuration, adapted to take better advantage of the new Burlington special armour. (US Army)*

influence on the new MBT programme. In July 1973, an American team visited the British Army research facility in Chobham, England, to examine a new special armour being developed. Previous US studies of special armours had concentrated on glass/ceramic composites layered inside steel armour. The new British armour, codenamed Burlington armour by the US Army but also known as Chobham armour referring to its place of origin, used a classified system of layered armour mounted on top of a basic steel armour shell. The new armour promised to offer exceptional protection against shaped charge warheads compared to ordinary steel warheads. The US representatives were so impressed with the demonstration that the Ballistics Research Laboratory (BRL) at Aberdeen Proving Ground began a crash program to develop an improved type for incorporation into the new MBT.

An added incentive to incorporate the special armour into the new MBT came in the wake of the October 1973 'Yom Kippur' war in the Middle East. The October war involved the largest tank-vs-tank fighting since World War 2. Moreover, the Israeli Army had used the latest US tank, the M60A1, while the Egyptians and Syrians had used the most modern Soviet export tank, the T-62. Clearly, a careful study of the lessons of this war were needed before the new US Army MBT design could be finalised.

## The XM-1 Tank

By the end of 1973, the new MBT design had been redesignated from XM815 to XM1 to symbolise that it was a radical new start in American tank design. The battlefield lessons of the October 1973 war proved critical in the XM1 programme for several reasons. One of the most striking tactical developments during the war was the extensive use of 9M14 Malyutka[1] guided anti-tank missiles

---

[1] Called AT-3 Sagger in the West.

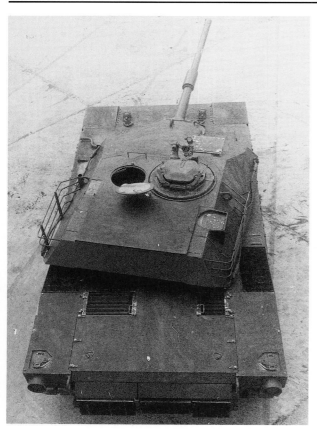

*An overhead view of the Chrysler XM1 validation phase tank. There were many changes to the turret of the tank between the validation phase and full scale engineering development. (US Army)*

and RPG-7 anti-tank rocket launchers by the Arab armies. The world press suggested that a large fraction of Israeli tank losses was caused by these weapons. Pundits claimed that they heralded the end of the tank's dominance on the modern battlefield, much as the longbow signalled the heyday of the armoured knight at Crécy in 1346. In fact, after-action studies of Israeli tank losses indicated that they had accounted for about 10 per cent of Israeli losses and that the primary cause of tank losses had been gunfire from other tanks. US studies also concluded that these infantry anti-tank weapons had been successful due to Israeli over-reliance on tank units without tactical co-ordination with mechanised infantry and artillery.

Even if the studies tended to downplay the infantry anti-armour threat, the desire to counter these types of weapons increased. US military analysts realised that the Egyptian use of new infantry anti-tank weapons was only a hint of what could be expected from the Soviet Army. There had been a steady increase in infantry anti-tank weapons in Soviet units since the 1950s, and it was likely that by the end of the 1970s, the most common anti-tank weapons on the battlefield would be various types of infantry anti-tank missiles and rockets, as well as related weapons fired from armoured vehicles such as the BMP-1's 73 mm projectile (a relative of the RPG-7 rocket) and missiles fired from helicopters and wheeled tank destroyers. These new anti-tank weapons had one potential weakness: they all used shaped charge warheads.[1] The new Burlington armour promised to render such weapons far less effective.

The lessons of the 1973 war also led to the decision to delete the 25 mm Bushmaster cannon. It was felt that tank crews were likely to engage light armoured vehicles with the main gun since an increasing number of these vehicles, notably the Soviet BMP-1 infantry vehicle, were being armed with anti-tank weapons. The most important lesson learned from the 1973 war was that the tank remained the dominant weapon on the modern battlefield, and, for that reason, the US Army was more committed than ever to adopt a new MBT.

The decision to incorporate the BRL version of Burlington armour on the XM1 forced General Motors and Chrysler to begin to redesign the armour layouts of their prototype vehicles. The validation-phase prototypes from General Motors and Chrysler were delivered to Aberdeen Proving Ground for the first series of developmental and operational trials from January till May 1976. The programme was temporarily delayed over the issue of whether the United States should adopt a common tank with Germany. The United States was pressing its NATO allies to standardise on weapon systems to reduce the logistics burden of each army having different types of weapons with incompatible parts, ammunition and fuel requirements.

---

[1] Shaped charge warheads are also called HEAT (high-explosive, anti-tank). In contrast to kinetic energy penetrators, such as the APFSDS round, the shaped charge penetrates armour by explosive force while kinetic energy rounds penetrate due to a combination of velocity and mass. HEAT warheads are preferred on low recoil, low velocity weapons such as infantry rockets and missiles since they depend on the warhead, not speed, to penetrate the target.

*The 2nd Armored Division was the first unit to re-equip with the M1 in 1982–83. One modification developed at Ft Hood was to cut back the rear skirt to prevent mud from building up in the drive sprocket. This was later adopted at the factory. The markings on the hull side identify this vehicle as belonging to the 'Hounds of Hell' 3-67 Armor. (Author)*

At the time, NATO was also considering the purchase of American E-3A Sentry AWACS aircraft. The German government strongly suggested that the US Army study the possibility of adopting the Leopard II for its new MBT requirement in view of its stated policy of weapons standardisation.

Much to the chagrin of US Army leaders, the Department of Defense agreed. A version of the Leopard II, called Leopard II AV, was sent to the United States where it underwent developmental trials in the autumn of 1976. The Leopard II was judged to have a superior fire control system, but inferior armour, ammunition compartmentalisation and gun traverse. One of the main problems was cost. FMC Corp. was interested in building the Leopard II in the United States if it was selected as the new US Army MBT. FMC concluded that Leopard II would cost 25 per cent more to build than either of the other XM1 prototypes. In January 1977, the US and FRG agreed that tank components, rather than whole tanks, should be standardised. The US pledged to seriously consider the adoption of the German Rheinmetall 120 mm gun on future production M1 tanks while the Germans pledged to consider employing the AGT-1500 turbine on their tanks.

At first, the US Army leaned towards issuing the contract to General Motors, favouring its design over the Chrysler design. But the army also wanted General Motors to modify its design to include the AGT-1500 turbine fitted to the Chrysler XM1 prototype. The Office of the Secretary of Defense (OSD) opposed the army approach, wishing to stretch the competitive aspects of the programme until after the two contractors had incorporated the engine changes as well as adding

*One of the initial low-rate-initial-production M1 Abrams tanks at Aberdeen Proving Ground shortly after their introduction into service. The XM1 FSED tanks can be distinguished by a more complicated ammunition feed on the commander's M2 .50cal. machine gun than the simple box feed on the LRIP tanks as seen here. This tank is finished in the four-colour MERDC camouflage scheme typical in the early 1980s. (Author)*

features to their turret designs that would allow the German 120 mm gun to be added at a later date. During this final phase of the bidding, Chrysler decided to subject their XM1 design to substantial revision in the hopes of pushing down unit cost, which they knew to be a key criteria of the OSD. The Chrysler team, under Dr. Philip Lett Jun., redesigned the special armour layout on the turret to take better advantage of the new technology. An extension was added between the gunner's sight and the commander's station to eliminate the separate commander's sight. Improvements and simplifications were also incorporated into the commander's cupola and the rangefinding system.

The army's insistence on the use of a turbine engine was due to its favourable experience with the switch to turbines on helicopters in the early 1960s. The army had found that turbines had a longer service life between major overhauls which significantly reduced lifetime maintenance costs. It is often forgotten that operation and maintenance costs on a tank significantly exceed its purchase price.

On 12 November 1976, Chrysler was declared winner of the full-scale engineering development (FSED) contract. The FSED phase of the programme required Chrysler to provide 11 pilot tanks for developmental and operational trials. The first tanks were delivered in February 1978. The tests uncovered problems with sand ingestion into the turbine engine which were resolved by improvement of the air filter seals. There were also problems with soil congestion on the drive sprockets which led to track shedding. This was solved by incorporating a simple mud scraper and a track retainer ring. PV 11 was subjected to survivability tests against various types of munitions after it was fully loaded with fuel and ammunition. The tests were successful and impressed the army with the considerable advances of the XM1 over the M60 in terms of crew survivability. Low initial rate production (LRIP) for the XM1 was authorised on 7 May 1979. These 110 LRIP tanks were used for the third and final phase of the operational trials with the 2-5th Cavalry at Ft Hood, Texas as well as extreme weather tests in the desert conditions at Yuma, Arizona, arctic conditions at the Alaska Cold Region Test Center, tropical conditions at Eglin AFB, Florida and electromagnetic and nuclear conditions at White Sands Missile Range.

The XM1 was accepted for full scale production and was type classified in February 1981. Initially, it was planned to name the tank after Gen. George C. Marshall, but the sentiment in the Armor branch was to name it after Creighton Abrams, a tank battalion commander with the 4th Armored Division in World War 2, head of the US Army in Vietnam in the later years of the Vietnam War, and a key supporter of the XM1 tank programme. As a result, the new tank officially became designated 105 mm Gun Tank M1 Abrams. The M1 was eventually produced at both the Lima, Ohio and Detroit, Michigan tank plants and production lasted until January 1985. During this period, Chrysler's defence division was sold off due to financial problems with the parent company, and became General Dynamics Land Systems Division (GDLS), ending Chrysler's long connection with US tank development.

The M1 Abrams entry into army service was troubled and controversial. The liberal mass media denounced the M1 tank as an expensive lemon, based mainly on test reports on the original pilot model XM1s. The M1 became a prime symbol of the Reagan defence build-up of the early 1980s, and so an especially obvious target for those opposed to the army modernisation effort. Further fuel was given to the critics by start-up problems at the M1 production plants, cost-overruns, and performance shortfalls. Many of the media charges were wildly off-base, and the excellent performance of the M1 Abrams battalions during the Reforger exercises in Germany in September 1982 heightened army enthusiasm for the new tank.[1]

**Improving the Breed**

From the outset, the US Army anticipated that the M1 Abrams would go through a series of evolutionary improvements. Indeed, the standard production M1 tanks had been configured to accept the German Rheinmetall 120 mm gun when necessary.

---

[1] For a more detailed description of the performance of the M1 Abrams at Reforger '82, see the original Osprey Vanguard 41 on the M1 Abrams tank.

The army attitude towards the upgunning was shaped by threat estimates of Soviet tank development as well as progress in 105 mm ammunition development. During the late 1970s and early 1980s, there had been a series of steady improvements in 105 mm kinetic energy penetrators which gave the US Army confidence that the 105 mm gun could penetrate Soviet main battle tanks at normal combat ranges of up to about 2,000 metres. Until the late 1970s, the standard US Army APFSDS round was the M735 which used a tungsten alloy penetrator with an initial velocity of 1501 m/sec. It could penetrate about 350 mm of steel armour at 2000 m. In 1979, the first depleted uranium (DU) penetrator round was introduced, the M774. This round had superior long range performance compared to the M735 and caused more damage on impact due to the pyrophoric reaction of uranium and steel. In 1983, the US Army began acquiring a significantly improved round, the M833, which used a longer and heavier DU penetrator. This round could reportedly penetrate 420 mm of steel armour inclined at 60° at 2000 metres. By way of comparison, the contemporary British 120 mm APFSDS had a penetration of 400 mm and the Soviet 125 mm projectile 450 mm according to unclassified reports.

The switch to APFSDS ammunition as the primary ammunition type for tank-vs-tank fighting was due to two factors. First, with the advent of special armours, both the British Chobham/Burlington armour and the Soviet K-combination (steel with ceramic inserts) armour, the new generation of tanks was disproportionately well protected against shaped charge warheads. For example, the M1 Abrams frontal armour was the equivalent of 350 mm of steel when attacked by

*An M1 Abrams of 2nd Platoon, Charlie Company, 2-64 Armor, 3rd Infantry Division (M) during the 1982 Reforger exercises. The 2-64 Armor was among the first units of US Army-Europe to convert to the Abrams and their superb performance during Reforger-'82 was widely viewed as a vindication of the Abrams design after a number of years of hostile press criticism. (Pierre Touzin)*

an APFSDS projectile, but it was the equivalent of 700 mm of steel versus a shaped charge warhead. Secondly, the kinetic energy rounds had better long range performance since they were generally less vulnerable to cross-wind effect due to their aerodynamic shape. In the 1960s and early 1970s, there had been considerable interest in guided anti-tank rounds for long range engagements. By the 1980s, however, guided rounds had fallen from favour due to vast improvements in fire control. The synergistic effects of advanced ballistic computers, laser rangefinders, wind sensors, barrel warp sensors and gun barrel thermal sleeves raised the accuracy of conventional tank projectiles to levels previously only obtainable with guided projectiles such as the US Army's 152 mm Shillelagh. The added attraction was that the conventional ammunition was only about 5 per cent of the cost of the guided munitions, even though the advanced fire control systems were quite expensive.

The US Army developed a longer barrelled, 60cal., version of the M68A1 105 mm gun for the M1 tank called the Enhanced 105 mm Gun (M24 Gun Tube) with an aim of replacing the older weapon when the added ballistic performance was needed to cope with newer Soviet tanks. This was a lower cost solution than replacing the M68A1 with the new 120 mm gun, since it would not require changes to the ammunition stowage system or the fire controls. But with the advent of even more advanced 105 mm ammunition such as the M900 105 mm APFSDS, there was even less reason for this change and it did not take place.

The second step in M1 Abrams evolution, called Block I, was the incorporation of the German Rheinmetall 120 mm gun. US studies of the gun concluded that it was overly complex and expensive by American engineering standards, so a version using fewer parts was developed, the M256. The prototype M1 with 120 mm gun was designated M1E1. The M1E1 also included other improvements including a new integrated nuclear-biological-chemical (NBC) protection system, an improved final drive/transmission and improved frontal armour. The first prototypes of the M1E1 were delivered in March 1981. Many of the changes on the M1E1 were viewed as suitable for the M1 as well, so in 1984 it was decided to manufacture an interim version, called IPM1 (Improved Product M1). The IPM1 tank included the suspension improvements, final drive, armour, and external stowage improvements of the M1E1, but neither the new NBC system nor 120 mm gun. A total of 894 IPM1s were built from October 1984 until May 1986, on top of the 2,374 basic M1s already produced by January 1985.

Consideration was also given to fitting the M1E1 with a Commander's Independent Thermal Viewer (CITV). Sometimes called a 'hunter-killer' sight, such a system had been incorporated in the original XM1 plans, but were dropped for cost reasons. On the M1, there was a single thermal imaging sight used by the gunner to acquire and engage targets. The commander received the same image as the gunner through an optical elbow. The CITV would allow the commander to search for new targets while the gunner was engaging a target. The CITV would have added substantially to the cost of the Block I tank, so it was deleted. However, a circular opening was provided in the roof so that the CITV could be retrofitted at a later date.

One of the longstanding controversies within the armour community was whether or not the

*The introduction of the M1 Abrams coincided with the introduction of the MILES laser simulation equipment. The canvas webbing at the base of the turret contains small laser sensors which sense whether the tank has been 'hit' by opposing forces, while the 9-round pyrotechnic simulator above the barrel simulates the firing of the gun. This is a tank of the 1st Cavalry Division at Ft Hood, Texas. (Author)*

*An M1 of 1-11 Armored Cavalry Regiment during Reforger-'83 in Germany. At the time, the unit was commanded by one of Creighton Abrams' sons, Col. John Abrams. During the war-games, the unit used improvised camouflage of mud smeared over the usual green camouflage colour, sometimes in elaborate patterns as seen here. (Pierre Touzin)*

M1 should be fitted with an auxiliary power unit (APU). The desire for an APU on the M1 stemmed from the fact that the turbine engine consumed fuel at a similar rate whether in idle or running full speed. This was a problem when the tank was in stationary overwatch in the field, as considerable fuel was consumed simply to keep the vehicle's electronics and sighting systems operating. An APU would help the fuel economy of the vehicle by powering the subsystems under such circumstances. The army examined both gasoline powered and gas-turbine APUs. The gas turbine APUs were generally favoured, since they could be incorporated within the engine compartment, under armour, by reducing some fuel tanks in size. Their main problem was that they were expensive, and they took up as much space as the amount of fuel they saved. Gasoline-powered generators, although much less expensive, were a potential fire hazard in the engine compartment. Instead, an external armoured box mounting was developed for a small diesel generator. A small number of these were purchased for trials and issued to units beginning in 1983.

The M1E1 was accepted for service in August 1984 and type classified as the 120 mm Gun Tank M1A1. The first production M1A1s came off the assembly lines at Detroit in December 1985. Priority was given to units stationed with the US Army in Europe (USAREUR). All USAREUR tank battalions received the M1A1 by the end of 1989 and all POMCUS facilities were equipped by June 1991.[1] In October 1988, the production plants began switching to a new variant of the M1A1, designated M1A1HA. The HA signified Heavy Armor, a reference to a special layer of depleted uranium mesh added to the armour package. There is no significant external difference between the M1A1 and M1A1HA. The new armour reportedly gives the M1A1HA the equivalent of 1300 mm of steel armour against shaped charge warheads and 600 mm against APFSDS, nearly

---

[1] POMCUS (Prepositioned Matériel in Unit Sets) refers to the large storage depots in Germany. In the event of war in Europe, US heavy divisions stationed in peacetime in the United States have their troops airlifted to Germany, and meet up with their equipment already in Europe in POMCUS.

*The Reforger exercises in Germany were held in the autumn or early winter. This M1 Abrams has a temporary winter camouflage pattern applied. The mast behind the radio antenna is a mounting for the 'whoopie light' used with the MILES simulator to indicate that a vehicle has been disabled. (Pierre Touzin)*

double the protection of the original M1 Abrams. This is the most effective armour package ever incorporated into a tank. The final production contract for US Army M1A1s was delivered in 1991 and production is expected to be completed by April 1993, bringing production to 4802 M1A1s.

The US Marine Corps had planned to begin purchasing 564 M1A1s in 1986, but the plans were delayed because of funding problems. The Marine Corps sponsored efforts to adapt the M1A1 to Marine requirements, including attachments for wading trunks to allow the tanks to wade ashore from amphibious assault craft. Rather than produce a separate Marine and Army M1A1 version, it was decided to merge the Marine features into later production M1A1s, resulting in a version called M1A1 Common Tank. A total of 60 M1A1HA tanks were loaned to the Marines from the US Army during Desert Storm. Delivery of 221 M1A1 Common Tanks to the Marine Corps began in November 1990 and was completed in 1992.

The M1 tank was considered for the Swiss Army's tank requirement, but the Leopard II was selected instead. The M1A1 and M1A2 were also alternatives in the British programme for a successor to the Challenger, but the Challenger II was selected. The M1 was put through trials in Saudi Arabia beginning in 1983 against the British Challenger and Brazilian Osorio. The Saudis decided to wait for the M1A2 version as related below. Egypt selected the M1A1 Abrams for its new MBT, with an aim towards co-producing 555 tanks locally. The first 25 were provided in kit form and assembled in 1991. Pakistan tested the M1 tank, but disagreements with the US over its nuclear programme have prevented acquisition of the Abrams tank.

With the Block I tank in production, attention shifted to the Block II programme. The Block II programme envisaged a radical redesign of the internal electronics of the vehicle. The US Army began to promote the 'Vetronics' idea, an integrated electronics system paralleling similar avionics systems being developed for aircraft. The Block II was envisaged as the first American digital tank. About a mile of wiring was taken out of the Block II. The electronics are based around digitial core architecture. The program originally envisioned an integrated command and control system (ICCS), now called IVIS which gives the commander a data display from which he can digitally transmit and receive reports via his SINCGARS radio. The Block II includes an inertial navigation system, codenamed POSNAV. Finally, the Block II incorporates the CITV commander's independent thermal viewer. The integration of the vetronics permits unique capabilities. For example, on encountering an enemy position, the commander can determine the range to the target using the tank's new $CO_2$ laser rangefinder. The computer architecture on the tank then calculates the precise location of the target using the laser data and the location data from the POSNAV. This information can then be rapidly and automatically transmitted as a 'call for fire' report to the divisional artillery.

Other improvements on the Block II include an improved commander's independent weapons station (ICWS) which uses new larger vision blocks, hardened against laser energy, and a simplified .50cal. machine gun mounting. The Block II tank, when accepted for service later in 1992, will be designated 120 mm Gun Tank M1A2. The first prototype was delivered in December 1990 and the first production vehicle is expected in November 1992. The initial production run of the M1A2 is expected to be small. Congress originally authorised the production of 62 tanks, but then reprogrammed funding to allow the construction of 120. In 1991, the Congress began steps to begin an M1 conversion plan which would rebuild M1s as M1A2s as part of a programme to keep at least one of the tank plants open after M1A2 production for the US Army ends in 1993.

Ironically, more M1A2s may be produced for foreign customers than for the US Army. Saudi Arabia selected the M1A2 as its new main battle tank. A total of 465 were ordered for delivery in 1993-96, and a further 235 tanks were later added to the order. The M1A2 was demonstrated to the Kuwaiti Army shortly after the Gulf War, and Sweden is examining the M1A2 and Leopard II after deciding against manufacturing their own MBT design.

The US Army has been undertaking the study of a new generation main battle tank since the mid 1980s. This programme has surfaced under various names including Armored Family of Vehicles (AFV), the Heavy Force Modernization Program (HFM) and currently Future Main Battle Tank

*Inside the turret of an M1 Abrams. To the right is the gunner's seat with his fire controls evident. To the extreme right is the commander's station: the cupola can be seen in the upper right of the photo. To the left is the breech of the M68A1 105 mm gun. (Author)*

*An interior view of the M1 looking forward in the loader's station showing the vehicle radio to the lower left and the gun breech to the right. (Author)*

(FMBT). For a time, consideration had been given to an evolutionary M1 Abrams tank, so this concept was labelled as Block III. In general, the US Congress has been reluctant to fund these programmes, arguing that the US Army needs a new generation of armoured artillery vehicles to replace the M109 sooner than it needs an M1 replacement. The excellent performance of the Abrams in the Gulf War as well as the breakup of the USSR in 1991 is likely to reduce funding for heavy force modernisation. In all likelihood, the main attention will be focused on new tank technology such as electro-magnetic and electrochemical guns, new projectiles, new powerplants and new armours which could be incorporated into a new tank at the end of the decade or early in the next century.

To support the development of new MBTs, the Abrams has served as the basis for a number of tank test-beds. Two of these are aimed at developing concepts and subcomponents for future MBTs. The Surrogate Research Vehicle developed in the early 1980s was designed to examine the feasibility of various novel turret configurations, particularly the electro-optical sensor layouts. The XM1E2 Close-Combat Test-Bed (CCTB) employs a reduced profile turret to further refine ideas about novel turret and hull layouts for a future MBT. The most recent experimental version is the CATTB (Components Advanced Technology Test-Bed) which will be equipped with the new XM291 120/140 mm ATAC Advanced Tank Armament Cannon System, the XM91 autoloader, the MTAS Multi-Target Acquisition Sensor, and the XAP-1000 AIPS Advanced Integrated Propulsion System. The new XM291, which can use either a 120 mm or 140 mm gun tube, was fitted to an M1 tank for trials in 1987-88.

## M1 Abrams Procurement[1]

|      | FY79 | FY80 | FY81 | FY82 | FY83 | FY84 | FY85 | FY86 | FY87 | FY88 | FY89 | FY90 | FY91 |
|------|------|------|------|------|------|------|------|------|------|------|------|------|------|
| Army | 110  | 309  | 569  | 665  | 855  | 840  | 840  | 790  | 810  | 689  | 679  | 481  | 225  |
| USMC |      |      |      |      |      |      |      |      |      |      |      | 66   | 149  |

[1] FY=Fiscal Year. This chart details how many Abrams tanks were funded in the US Army budget. Actual calendar year production differs: 813 (1984); 705 (1985); 755 (1986); 947 (1987); 784 (1988); 720 (1989).

In view of the fact that the M1 and M1A1 are likely to remain the US Army's primary main battle tank for the foreseeable future, the US Army initiated a 'Material change program' in 1990 to gradually upgrade the existing fleet. The programme is divided into eight blocks, designated Blocks A to H. Four of these Block improvements have already been started and are mainly aimed at minor engineering corrections, fire safety improvements and many small depot upgrades. In June 1992, the army announced plans to upgrade about 400 early-production M1A1 tanks which had served in Operation Desert Storm. The vehicles would have the heavy armour packages inserted as well as receiving the digital electronics architecture of the M1A2. However, they would not be fitted with the CITV or some other M1A2 features. If the programme goes ahead, these vehicles will be designated M1A1-D. The two final upgrades, which have not yet been formally approved or funded, include Block G which would add a new armour package to the M1A2 beginning in January 1995, and Block H which envisages upgrading M1A1s with M1A2 features and other improvements, and retrofitting a portion of the M1 fleet with some M1A2 features such as the 120 mm gun. The Block H programme will be heavily

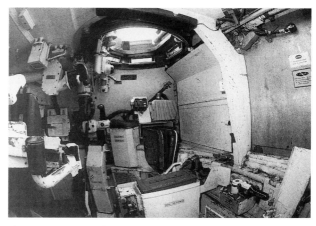

*An interior view of the M1 looking from the loader's station in the left side of the turret over towards the commander's station. To the right are the blast doors for the ammunition. To the left is the gun breech. (Author)*

dependent on the level of US Army funding as well as congressional attitudes towards the army's future role.

There are a large number of programmes examining possible improvements that might be incorporated in future M1 modification programmes. The MSGL Multisalvo Smoke Grenade Launcher recognises the role that smoke can play in increasing the survivability of the M1A1 Abrams, especially if faced with enemy tanks with thermal sights.

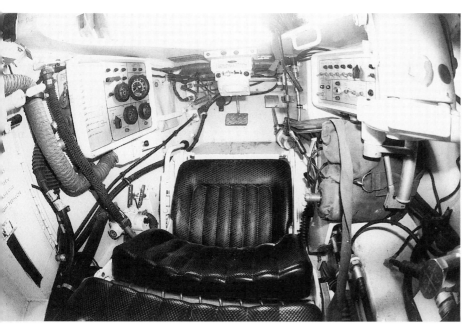

*A view of the driver's station in the M1. The driver sits in an almost prone position when the tank is buttoned up. His main control is a set of hand-controls reminiscent of those on a snow-mobile or motorcycle. (Author)*

*The sharp end of the M1 Abrams' weapon system is the armour-piercing fin-stabilised discarding-sabot (APFSDS) kinetic energy penetrator, more commonly nicknamed 'sabot' ammunition by US tankers. In this remarkable view, the sabots are peeling away from the long-rod penetrator. (Alliant Techsystems)*

# OPERATIONAL HISTORY

Unlike the current grenade launchers which need to be reloaded after each use, the new system would provide several bursts of smoke grenades between reloading. Connected with this programme is the XM81 MM/IR screening grenade. This smoke grenade is designed to protect the tank against enemy weapons using millimetre wave radar sensors or thermal imaging sights. It is extremely likely that a BCVI (Battlefield Combat Vehicle Identification) system will be adopted in the next few years, especially in light of the fratricide problem during Operation Desert Storm. There is a four-stage programme under way to examine alternative BCVI technologies. The LWR (Laser Warning Receiver) programme is intended as an adjunct to other efforts to defeat enemy guided anti-armour weapons. A number of new anti-tank weapons, including Soviet tank gun launched missiles such as Bastion and Svir are laser guided. The LWR would warn the tank crew when they were being lased, so that countermeasures could be undertaken, such as launching smoke grenades.

The M1 Abrams tank was first committed to combat in 1991 during Operation Desert Storm. During the war, the M1A1 Abrams was credited with destroying over 2000 Iraqi tanks, with not a single Abrams tank being destroyed by hostile tank fire.

The first heavy unit deployed to Saudi Arabia in August 1990, the 24th Infantry Division (Mechanized), was still equipped with M1 and IPM1 Abrams tanks. As from November 1990 there were 580 M1/IPM1s in Saudi Arabia and only 123 of the newer M1A1 tanks. By the end of 1990, when it was becoming increasingly evident that ground combat was likely, the US Army began a roll-over programme to re-equip all heavy units in Saudi Arabia with the improved M1A1 tanks.

The decision to re-equip all heavy divisions in Saudi Arabia with the M1A1 Abrams was due to a variety of factors. It was widely presumed that the Iraqi Army would use chemical weapons, and the M1A1, with its integral NBC system, was better suited to operating in a chemical environment than the earlier M1 Abrams. The M1A1, particularly the M1A1HA variants, had substantial improvements in armour protection, reliability and firepower, all critical ingredients in tank fighting. As there would not be enough M1A1HA tanks readily available from the factory, a modification effort was begun to upgrade 865 earlier M1A1s with M1A1HA features. This effort included armour improvements, fire control modifications, new engine fire shields, an NBC system heat exchanger, repainting with desert sand CARC paint and a host of other changes.

One of the more curious improvements added to about 800 tanks during the roll-over programme was the Sleep Support System (SSS). Since it was presumed that in a chemically contaminated environment the crew would have to sleep in the tank, the SSS was developed to allow two crewmen at a time to sleep within the confines of the tank. It consisted of a set of hammocks and attachment points set up in the loader's and driver's station. Another unusual programme was the laser protection programme. There had been reports that the Iraqi Army had used lasers to blind Iranian forces during the Gulf War in 1987-88. A special insert filter was developed for the standard protective goggles issued to troops. It is unclear how many were actually issued.

An important innovation for desert operations was the Small Light Weight GPS Receiver, nicknamed Slugger. This system receives signals from overhead GPS Global Positioning System navigation satellites and provides the tank crew with precise location data. Not all tanks in theatre were equipped, but the success of GPS led to a move to equip all M1A1 tanks in the future with the device. As an alternative and back-up to GPS, tank units also received LORAN receivers. LORAN is commonly used at sea for navigation, and the US Army found that many tankers were already familiar with the system from deep sea fishing vacations.

Another major effort was directed towards improving the readiness rate of the M1s already in Saudi Arabia. By the end of 1990, the readiness rate was about 89 per cent, the main source of problems being high engine usage and associated sand ingestion problems, drive sprocket wear, gunner's primary sight parts shortages, and track wear. A significant portion of the tanks in Saudi Arabia, especially those from stateside units such as 24th Infantry and 1st Cavalry, were very old production vehicles. It was expected that once the roll-over was completed, this would aid the readiness rates. The new T-158 track was sent to Saudi Arabia and by the outbreak of the war, about 20 per cent of the M1A1s had this in place of the older T-156 track. By February 1991, the US Army had 1,956 M1A1 Abrams tanks (1,223 M1A1HA, 733 M1A1)[1] deployed with units in

*An IPM1 undergoes repairs by an M88 recovery vehicle at the National Training Center at Ft Irwin, California. Most US tankers feel that the realistic war-games at NTC were the prime reason for the outstanding performance by US armour units during Desert Storm. (Author)*

*An AGT-1500 gas turbine powerpack is pulled from an IPM1 Abrams at NTC by an M88 recovery vehicle. Although the turbine has proved to have prodigious fuel consumption, its greater reliability compared to diesels has lived up to expectations. (Author)*

optically tracked by means of a flare at the rear of the missile. These new jammers work by projecting an intense beam of light at a similar frequency which confuses the missile tracker and causes the missile to crash away from the targeted tank. The US Army programme was officially labelled as the AN/VLQ-8A Missile Countermeasure Device (MCD), although it is also called Hardhat by its developer, Loral Electro-Optical Systems. Contracts for the first batch of these were issued in January 1991, but none were available in time for use in the Gulf.

The three armoured and two mechanised infantry divisions deployed to the Gulf by the United States during Operation Desert Storm accounted for the majority of M1A1 tanks serving in the theatre. The organisation of US Army tank battalions at the time of the war conisted of a HQ company with two M1A1 tanks and four tank companies, each with 14 M1A1 tanks for a total of 55 tanks per battalion. The organisational structure of armoured and mechanized infantry divisions are similar, the major difference being that an armoured division has six armour and four mechanised infantry battalions while the mechanised infantry division has five armour and five mechanised infantry battalions. As it was decided not to use National Guard 'round-out' brigades to bring up to strength several of the divisions, brigades from other regular divisions not deployed to Saudi Arabia were used for this purpose. One brigade from the 2nd Armored Division with two M1A1 tank battalions was attached to the Marine Corps forces deployed to the Gulf to augment the firepower of their M60A1s.

The other major formation equipped with M1A1 tanks deployed in the KTO (Kuwait Theatre of Operations) were the two armoured cavalry regiments. These were split, one serving in each of the US Army Corps. These regiments consist of three armoured cavalry squadrons each with three armoured cavalry troops (nine tanks and 12 M3A2 Bradleys each) and a tank company (14 tanks). Thus each squadron has a total of 41 tanks and 38 M3A2 Bradleys and each regiment 116 M1A1 Abrams and 132 Bradleys.

Saudi Arabia, with some units exchanging tanks in the last weeks before the ground war began. A further 528 tanks were sent to Saudi Arabia, but kept in reserve stock. All units deployed for combat were equipped with M1A1 or M1A1HA tanks.

Several programmes were under way which were completed too late to be used in combat. US Army intelligence had picked up evidence of the Iraqi Army deploying anti-tank missile jammers, nicknamed 'dazzlers'. Wire-guided anti-tank missiles, by far the most common type of ATGM, are

[1] Other US Army sources indicate other numbers: 1178 M1A1 and 594 M1A1HA. The discrepancy between the two numbers is unexplained.

During the ground war, some battalions were

cross-attached to form battalion task forces. An example would be for a tank battalion to trade one of its tank companies with a Bradley infantry battalion, getting a Bradley company in exchange, and forming a battalion task force in the process.

The US Marine Corps deployed five tank battalions to the South West Asia theatre. The Marines had 60 M1A1HAs on loan from the army, plus 16 new M1A1 Common Tanks. The 2nd Tank Battalion was re-organised under US Army organisation tables with 54 M1A1 tanks, rather than in the under the standard Marine tables which have 70 tanks per battalion. The remaining Marine M1A1 tanks were used by the 4th Tank Battalion.

*An IPM1 on the prowl in the Mojave desert at NTC. The IPM1 has the longer, uparmoured turret of the M1A1, but is otherwise very difficult to distinguish from the basic M1 Abrams. The IPM1s at NTC are finished in a non-standard camouflage pattern of sand, field drab and black. (Author)*

# DESERT STORM: M1A1 TANK BATTALIONS

US Army 7th Corps
**2nd Armored Cavalry Regiment**
(1-2, 2-2, 3-2 Cavalry1)
**1st Cavalry Division**
(1st Brigade: 3-32 Armor, 2-8 Cavalry; 2nd Brigade: 1-32 Armor, 1-5 Cavalry, 1-8 Cavalry)
**1st 'Old Ironsides' Armored Division**
(1st Brigade: 1-37 Armor, 4-66 Armor; 2nd Brigade: 1-35, 2-70, 4-70 Armor; 3rd Phantom Brigade (from 3rd Infantry Division: 4-66th Armor)
**3rd Armored Division**
(4-7 Cavalry; 1st Brigade: 4-32 Armor; 2nd Brigade: 3-8, 4-8 Cavalry; 3rd Brigade: 2-67, 4-67 Armor)

*An M1 Abrams at Ft Sill, Oklahoma in 1991. This vehicle might be mistaken for an IPM1 due to the rear stowage basket and the absence of the retainer ring on the drive sprocket. In fact, since the late 1980s, many M1s have been retrofitted with the stowage rack due to general disgruntlement with the paucity of external stowage on the basic M1. This vehicle has the original short M1 turret which is evident by noting the distance between the tow-cable attachment and the front turret armour joint. On the IPM1, the distance is about 9 ins. greater. (Author)*

**1st Infantry Division**
(Mechanized) (1st Brigade: 1-34, 2-34 Armor; 2nd Brigade: 3-37, 4-37 Armor; 3rd Brigade [from 2nd Armored Div. FWD]: 2-66, 3-66 Armor)

### US Army 18th Airborne Corps
**3rd Armored Cavalry Regiment,**
(1-3, 2-3, 3-3 Cavalry)
**24th Infantry Division**
(Mechanized) (1st Brigade: 4-64 Armor; 2nd Brigade: 1-64, 3-69 Armor; 197th Infantry Brigade: 2-69 Armor)

### 1st Marine Expeditionary Force
**1st Tiger Brigade**
(US Army) (from 2nd Armored Division: 1-67, 3-67 Armor)
**US Marine Corps,**
(2nd, 4th Tank Battalion)

### G-Day
The Iraqi Army in Kuwait and in the neighbouring sectors of Iraq had eight armoured, four mechanised and 31 infantry divisions with a total of about 4110 tanks and 2570 armoured infantry vehicles. The 12 heavy divisions included the bulk of the Iraqi tank force; each infantry division usually had a tank battalion with about 35 older tanks, normally Polish T-55, Chinese Type 59 or Type 69 tanks. The best Iraqi tank equipment – Soviet, Polish or Czechoslovak-manufactured T-72 and T-72M tanks – equipped the Republican Guards divisions. These units were in the third strategic echelon of the Iraqi forces, generally to the north and west of Kuwait. Some selected second-echelon divisions such as the 3rd Saladin Armored

Division also received T-72s for at least one of their brigades.

At the outset of the ground campaign, the US Air Force claimed that about 40 per cent of the tanks and 35 per cent of the other armoured vehicles had been destroyed by air attack. It would later become apparent that these claims were excessive. In Kuwait itself, the US Marines found that only 10-15 per cent of the destroyed armoured vehicles had been knocked out by air attack, primarily by attack helicopters and to a lesser extent by A-10s and other aircraft firing Maverick missiles. In the areas where the US Army operated, air kills varied from about 15-25 per cent. The Republican Guards units in north-western Kuwait and neighboring Iraq, were singled out for the heaviest air strikes. The US 3rd Armored Division estimated that 34 per cent of the tanks and 23 per cent of other armoured vehicles of the three Iraqi heavy divisions in their sector had been destroyed by air attack before the outbreak of the ground campaign. Although the air attacks did not have as great an effect on matériel as had been estimated, their impact on Iraqi morale, command and control and logistics preparation was devastating.

The heaviest elements of the US Army equipped with the M1A1 tanks were in VII Corps

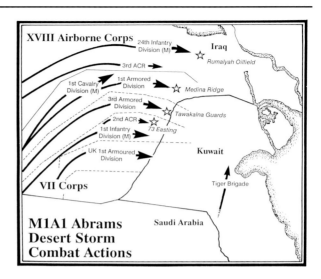

*Another example of a modernised M1 Abrams with the rear turret basket, this time Ballbuster, the command tank of B Troop, 1-11 ACR at Fulda, Germany in 1988. The spacing between the front tow cable attachment and the turret armour joint on the right turret side is different than on the left, and more difficult to gauge without reference to other distinguishing features of the IPM1, such as the front gun mantlet. This vehicle is unusual in that it is painted in overall olive drab for a change of command ceremony, not the usual forest green. (Stephen Sewell)*

and XVIII Airborne Corps. They were committed to the 'Hail Mary' sweep through the deserts to the west of Kuwait, with an aim towards blocking Iraqi withdrawal into the Euphrates River valley and destroying the Republican Guards divisions.

The main problem facing US armour battalions in the first hours of G-Day was posed by Iraqi minefields and sand berms along the frontier. Aside from the use of dedicated combat engineer equipment, the army tank units had been issued with Battalion Countermine Sets (BCS). The principal elements of BCS equipment were mine ploughs, with three sets issued to each company. These were patterned off Soviet mine rakes and basically pushed aside any anti-tank mine encountered. In addition, each battalion received six mine roller mounting kits and each company a single roller. The rollers were considerably heavier than the ploughs (11 tons vs 4 tons) and so were generally shed once the initial breaching operations were completed. Each tank company was also issued with a single Cleared Lanes Marking System (CLAMS) which was attached to the rear of the tank to mark a path through the cleared minefields. The CLAMS dropped small incandescent flares at predetermined intervals.

The breaching operation was viewed with some trepidation by US Army planners, as there was fear that the Iraqis would saturate the attack lanes with artillery fire and chemical weapons. The crews of all Abrams tanks participating in the initial

*A rear view of an M1A1 of 4th Platoon, B Troop, 1-11 ACR. Since the early 1980s, US tank units have adopted the practice of mounting metal plates at the turret rear with geometric symbols to indicate unit affiliation. This is to make it easier for unit commanders to visually identify tanks in their unit. (Pierre Touzin)*

The breaching was accomplished without the feared Iraqi chemical attacks. Iraqi artillery proved largely ineffective due to the damage caused by weeks of air strikes, and the disruption of Iraqi command nets. When the Iraqi artillery did fire, it was quickly silenced by US Army counter-battery fire, most often the 'steel rain' of MLRS Multiple Launch Rocket System. Once through the initial Iraqi barrier network, the objective of the US Army heavy divisions was to plunge as deeply and as rapidly as possible into Iraq. It was a sight rarely seen in modern warfare: entire divisions racing across the flat open desert. American tankers recalled that the sight of wave after wave of equipment stretching to the horizon was 'breathtaking'. A single division, such as the 24th Infantry, consisted of 1793 tracked combat vehicles, 6566 wheeled vehicles and 94 helicopters, all moving on a broad front across the desert.

assaults were dressed from the outset in chemical protective equipment. The most controversial incident in the breaching operation was the decision by the 1st Infantry Division to use the mine ploughs on the Abrams tanks more aggressively. Once paths were cleared, the tanks with the mine plooughs attacked Iraqi trench lines, using the ploughs to bury any Iraqi infantrymen who did not flee or surrender. Total Iraqi casualties in this operation are not clear, but it caused a fuss in the press after the war.

Resistance in the first two days of fighting in the Iraqi desert was modest, as there were no large Iraqi armour concentrations in the area. There were several intense fights, but the outcomes were lopsided. For example, 4-64 Armor of the 24th Infantry Division encountered stiff resistance when it reached Battle Position 102 east of An Nasiriyah. The Iraqi forces there consisted of an Iraqi commando brigade backed by air defence artillery units and tanks. The positions were

*An M1A1 Abrams of C Troop, 1-3 ACR in Germany during autumn Reforger exercises. The blue rectangle on the turret front is a temporary war-game marking, blue for friendly forces, orange for opposing forces. With the introduction of the M1A1 Abrams in Europe in the late 1980s, the US Army began switching to the common three-colour NATO camouflage scheme. (Pierre Touzin)*

*An M1A1HA with mine ploughs of the 24th Infantry Division (M) at Ft Stewart, Georgia, in 1992 after their return from the Gulf. The 24th Infantry Div. is part of the US Army contingency corps, and so retains the sand desert camouflage even when stationed back in the USA. This tank is marked in the same fashion as during Desert Storm. (Author)*

overwhelmed though a few M1A1 Abrams suffered damage from RPG rocket grenades. On 25 February, the 2nd Armored Cavalry Regiment in the vanguard of the VII Corps attack destroyed a reinforced mechanised infantry battalion with T-55s and MT-LBs of the Iraqi 12th Armored Division near Phase Line Blacktop. It was a hint of more to come. The heaviest tank fighting of the war began on G+3.

By the third day of the fighting, 26 February 1991, VII Corps began to swing back east to engage the Iraqi armour formations in north-western Kuwait. The main Iraqi armour forces in the area west of Wadi al Batin was the Republican Guards Forces Command (RGFC) 3rd Tawakalna and 2nd Medinah Divisions, and the 12th Mechanized and 12th Armored Divisions.[1] Lead elements of the US Army's 1st and 3rd Armored Divisions and the 2nd ACR began engaging the Iraqi armoured formations around midday.

The 2nd ACR was wedged between 3rd Armored to its north and the British 1st Armoured Division to its south. The first fighting on 26 February began around 0700 hrs against a small Iraqi security force of 10 T-72 tanks and BMPs near 60 Easting (Easting was slang for north-south longitudinal lines on US Army maps). Air strikes were called in. By 0900 hrs, the area was enveloped by a shamal dust storm which grounded the 2nd ACR's helicopters and reduced visibility to 200-1,400 metres. For the remainder of the morning the M1A1s and Bradleys of the regiment continued to encounter isolated Iraqi tank formations in revetments. In many cases, the Iraqi positions were brought under fire from long range before they could respond. As the Bradley and Abrams troops began reaching the 70 Easting line in the early afternoon, they began to encounter very determined resistance by T-72 tanks and BMP infantry vehicles of 50 Brigade, 12th Armored Division, beginning what became known as the 'Battle for 73 Easting'. To the north, Ghost Troop crossed 73 Easting around 1620 hrs and destroyed 13 T-72s and 13 BMPs positioned on a reverse slope in a shallow wadi. 'All I could see were things burning for 360 degrees, nothing but action.' Another trooper recalled, 'Wave upon wave of tanks and infantry would come at Ghost only to be destroyed.' One Bradley was knocked out by T-72 tank fire. Eagle Troop in the centre began its main engagement around 1607 hrs and in a vicious 23 minute engagement destroyed 28 Iraqi tanks and about 50 other vehicles. To the south, Iron Troop

---

[1] The formal names of these two RGFC divisions are Tawakalna ala Allah, and Medinah Manarawah.

*An interior view of the M1A1 tank looking from the loader's station towards the gunner and tank commander. The interior of the M1A1 is very similar to the M1 version, with the obvious exception of the gun breech. (GDLS)*

encountered the Iraqi positions around 1830 hrs, knocking out nine T-72s and four BMPs. 'The annihilation of this Iraqi armor battalion continued as the Troop found itself surrounded by burning hulls and exploding ammo bunkers. The unforgettable odor of burning diesel, melting metal and plastics, expended munitions and anything else that happened to be burning in the bunkers, hung heavy in the air. For a moment there was an abrupt calm. An occasional shot from a TOW or a tank kept us alert. The scouts were told to continue their advance as the tanks held the line and overwatched. The report of advancing T-72s from the east told us the battle wasn't over.'

The encounters all took place under dismal weather conditions, with the dust storms and low cloud cover making it very difficult to see even with thermal sights. But it was an unequal contest. 'At 2,100 meters the inferior T-72 didn't stand a chance against the M1A1 Abrams. The depleted uranium long rod penetrator from the sabot round passed through the T-72s like a hot knife through butter. The TOW missiles (from the Bradleys) also had no problem and the counter-attack was squelched like a match in a cup of water.' The tankers later concluded that the fighting had been so violent because units of the Iraqi 12th Armored Division were trying to escape northward. As night fell, the T-72s continued to

# M1A1 Abrams, A Co., 1-37 Armor, 1st Armd. Div., KTO, January 1991

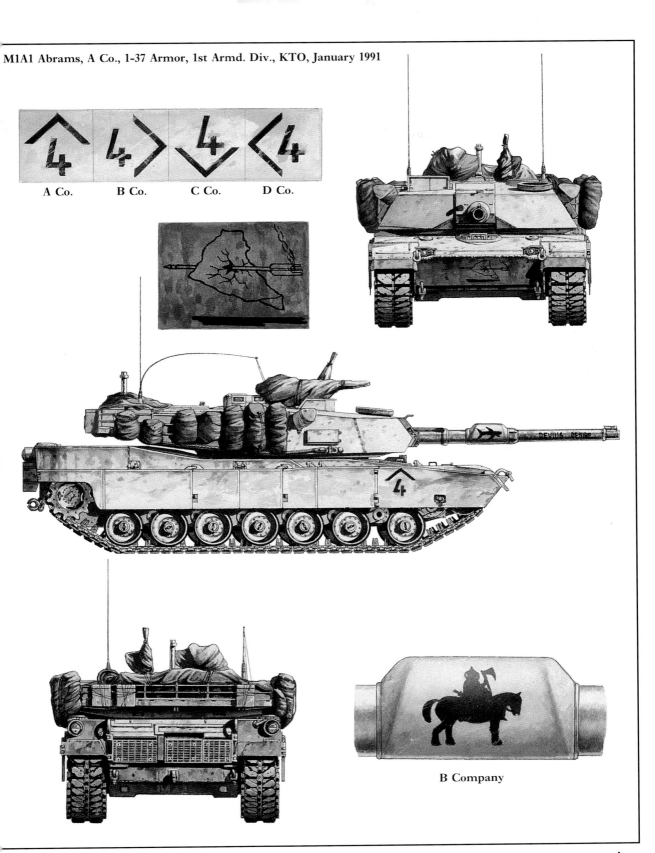

A Co.  B Co.  C Co.  D Co.

B Company

M1A1 Abrams, 3-66th Armor, 1st Inf. Div., KTO, January 1991

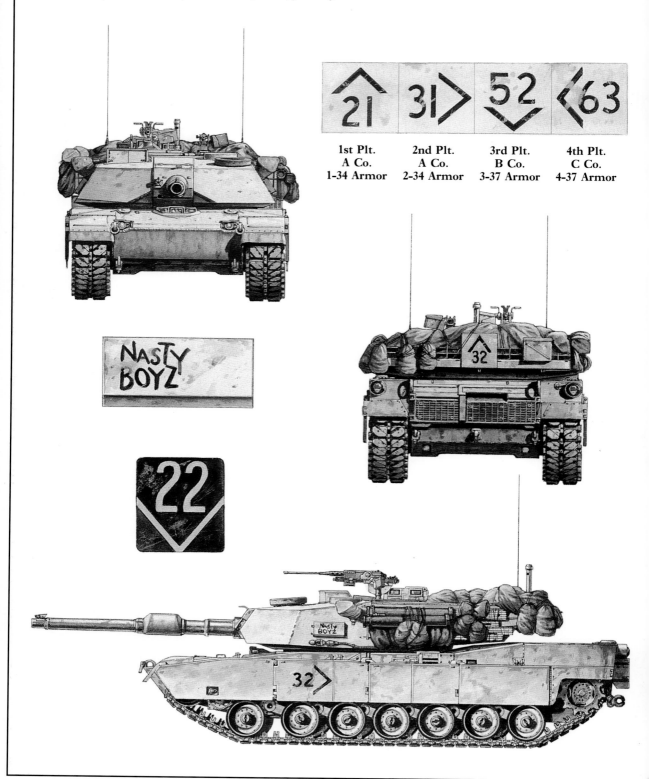

M1A1 Abrams, 3-8 Cavalry, 3rd Armd. Div., K1O, February 1991

# M1A1HA ABRAMS

*H Troop, 3rd Armored Cavalry Regiment, Operation Desert Storm, February 1991*

## SPECIFICATIONS

**Crew:** 4
**Combat weight:** 62.6 tons
**Power-to-weight ratio:** 24.0 Hp/ton
**Hull length:** 26.0 ft
**Overall length:** 32.3 ft
**Width:** 12 ft (with side skirts), 11.4 ft (side skirts removed)
**Engine:** Textron Lycoming AGT 1500 gas turbine, 1500 Hp
**Transmission:** Allison X1100-3B hydrokinetic, 4 forward 2 reverse
**Fuel capacity:** 504.5 US gallons
**Max. speed (road):** 41.7 mph
**Max. speed (cross-country):** 30 mph
**Best cruising speed:** 25 mph
**Max. range:** 275 miles at cruising speed
**Fuel consumption:** 1.83 gallons per mile
**Fording depth:** 4.0 ft (unprepared), 7.8 ft (prepared)
**Armament:** M256 120 mm cannon
**Main gun ammunition:** M829 APFSDS (Armour-piercing, fin-stabilised, discarding sabot); M830 HEAT-MP (high-explosive, anti-tank, multi-purpose)
**Muzzle velocity:** 5500 ft/sec (APFSDS), 3735 ft/sec (HEAT-MP)
**Max. effective range:** 3500 m (APFSDS), 3000 m (HEAT-MP)
**Stowed main gun rounds:** 40
**Gun depression/elevation:** −10 degrees/ +20 degrees

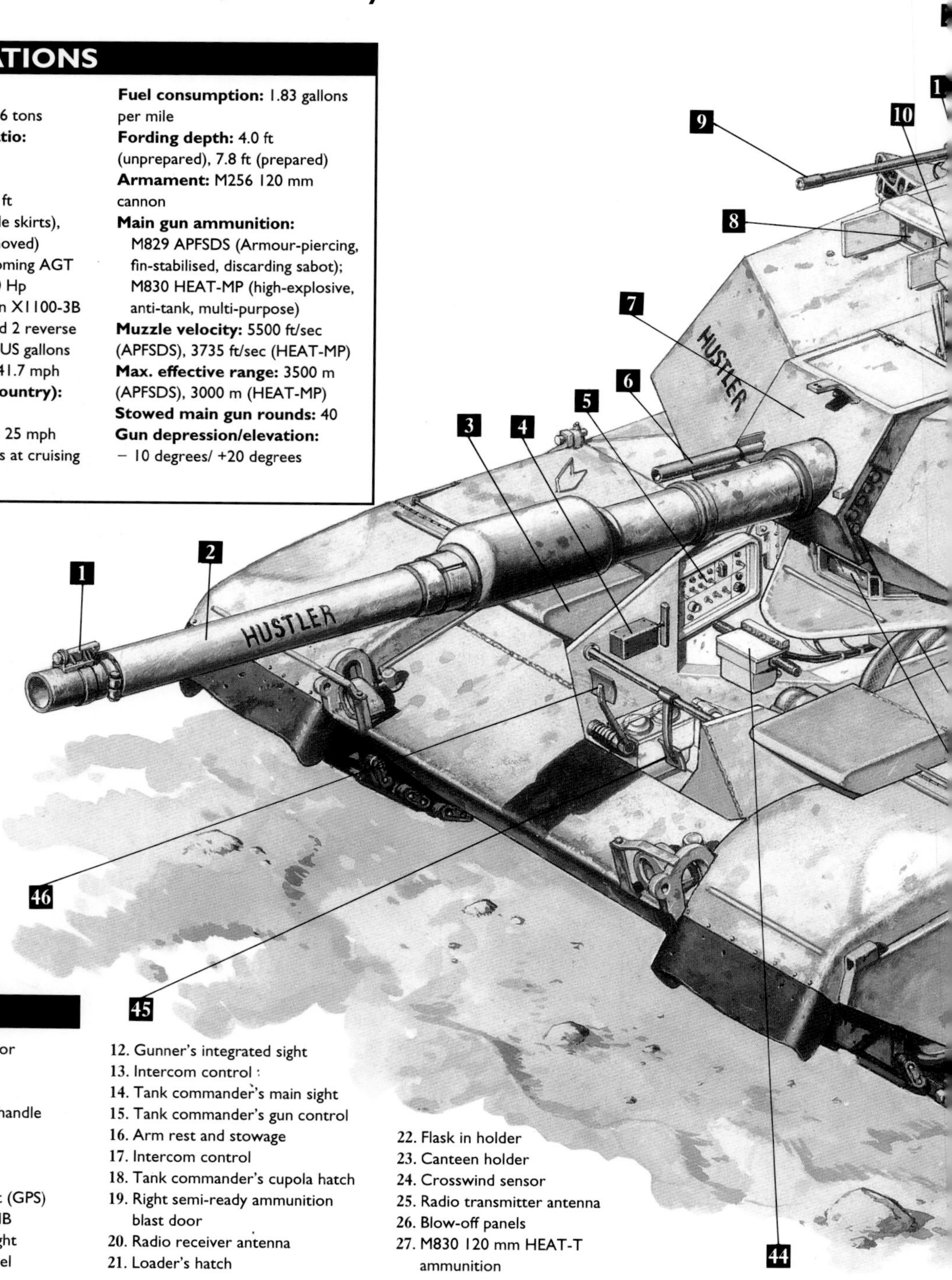

## KEY

1. Muzzle reference sensor
2. M256 120 mm gun
3. Fuel tank
4. Parking brake release handle
5. Driver's master panel
6. Co-axial machine gun
7. Gunshield
8. Gunner's primary sight (GPS)
9. M2 .50 cal Browning HB
10. Gunner's telescopic sight
11. Computer control panel
12. Gunner's integrated sight
13. Intercom control
14. Tank commander's main sight
15. Tank commander's gun control
16. Arm rest and stowage
17. Intercom control
18. Tank commander's cupola hatch
19. Right semi-ready ammunition blast door
20. Radio receiver antenna
21. Loader's hatch
22. Flask in holder
23. Canteen holder
24. Crosswind sensor
25. Radio transmitter antenna
26. Blow-off panels
27. M830 120 mm HEAT-T ammunition

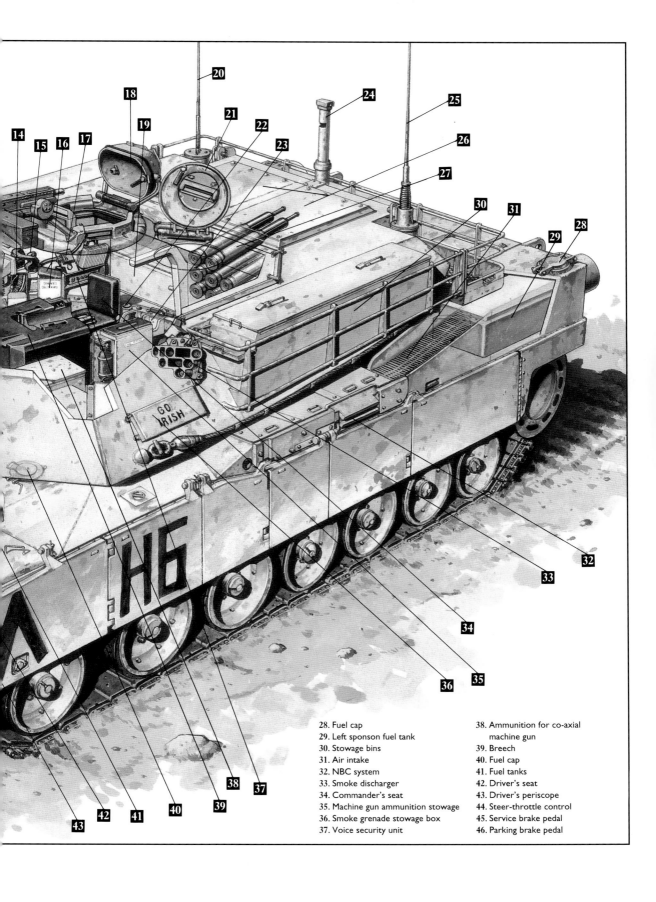

28. Fuel cap
29. Left sponson fuel tank
30. Stowage bins
31. Air intake
32. NBC system
33. Smoke discharger
34. Commander's seat
35. Machine gun ammunition stowage
36. Smoke grenade stowage box
37. Voice security unit
38. Ammunition for co-axial machine gun
39. Breech
40. Fuel cap
41. Fuel tanks
42. Driver's seat
43. Driver's periscope
44. Steer-throttle control
45. Service brake pedal
46. Parking brake pedal

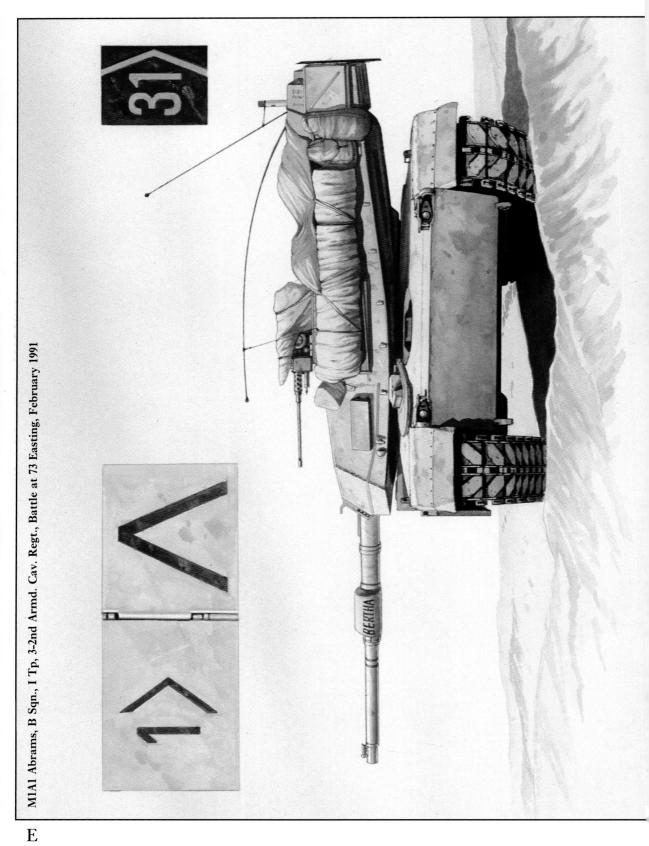

M1A1 Abrams, B Sqn., I Tp, 3-2nd Armd. Cav. Regt., Battle at 73 Easting, February 1991

E

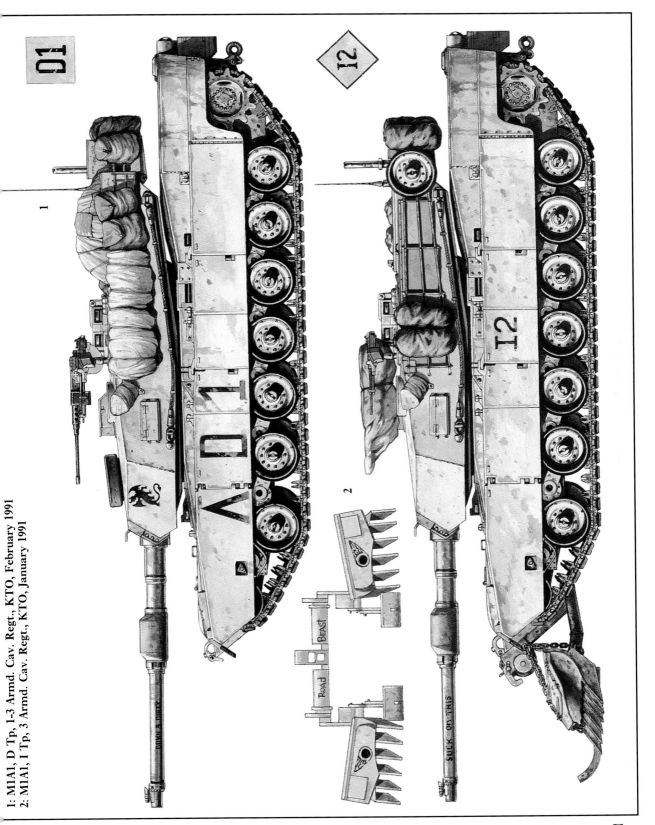

1: M1A1, D Tp, 1-3 Armd. Cav. Regt., KTO, February 1991
2: M1A1, I Tp, 3 Armd. Cav. Regt., KTO, January 1991

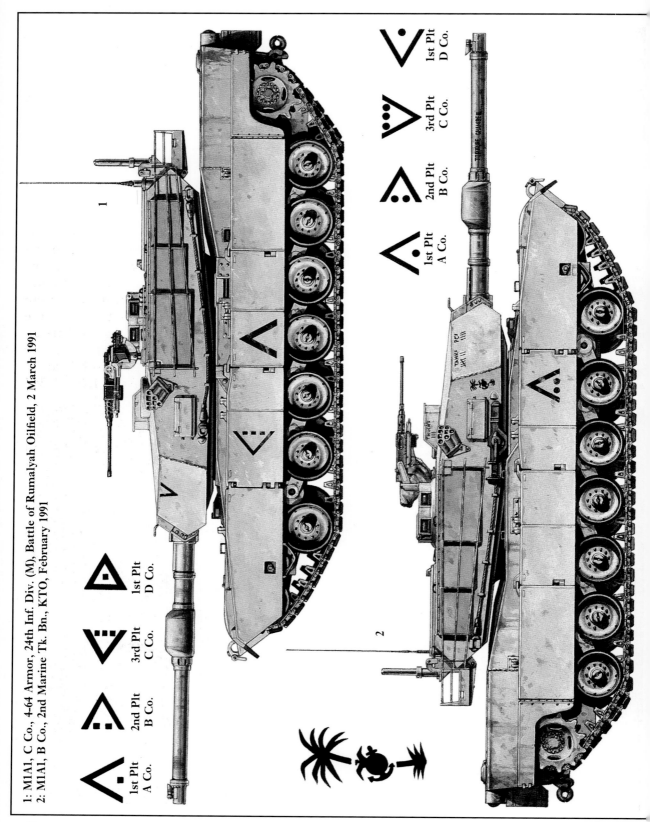

1: M1A1, C Co., 4-64 Armor, 24th Inf. Div. (M), Battle of Rumalyah Oilfield, 2 March 1991
2: M1A1, B Co., 2nd Marine Tk. Bn., KTO, February 1991

*The most common element of the Battalion Countermine Set (BCS) is the mine plough. This system is patterned on Soviet mine rakes. It scoops up anti-tank mines and pushes them to the side. (Author)*

attack, '(Iraqi) tanks were coming over the hill like there's no tomorrow...they were fighting for their lives, trying to get out,' recalled one trooper.

The scenes were repeated north of the 2nd ACR's positions. Bradley's of the 3rd Armored Division's 4/7 Cavalry became tangled in a firefight with T-72 tanks of the Tawakalna Division in the early morning and four Bradleys were hit. The Iraqis had prepared an ambush position with 35 tanks and supporting BMPs. They were eliminated by an Apache attack helicopter battalion before the weather closed in. The division continued to move forward with sporadic contact against mixed assortments of T-55s, T-62s and T-72s. The 3rd Brigade's armour battalions were brought forward from the divisional reserve when it was realised that the main Iraqi armour concentrations were being encountered. As it would transpire, the 3rd Armored Division had run into the heaviest concentration of Iraqi armor that would be encountered during the war. West of Wadi al Batin was the main body of the RGFC 3rd

*The M1A1 Common Tank was developed to satisfy Marine Corps requirements for landing their M1A1 Abrams tanks from amphibious landing craft. This tank was taking part in trials in 1989 prior to the entry of the M1A1 into Marine service. (US Marine Corps)*

*An interior view of the M1A2 Abrams. One of the most significant differences is in the area of the commander's station which has a new display for the CITV as well as a data display linked to the vehicle's new digital electronics. (GDLS)*

Tawakalna Division. East of the wadi was the 10th and 12th Armored Divisions. Some of the Iraqi tank formations were in well prepared defensive positions, while other Iraqi tank battalions were encountered out in the open, some attacking, others moving north in an attempt to escape. During the fighting on 26-27 February, the 3rd Armored destroyed 374 Iraqi tanks and 404 infantry vehicles in a series of intense firefights. Of these, Apache attack helicopters accounted for 32 of the tanks and 160 other vehicles. What had been remarkable was the accuracy of the tank guns: only 774 120 mm tank rounds were fired. The Bradleys had played a critical role in the armour fighting, expending about 10,100 rounds of 25 mm ammunition and 101 TOWs.

Further to the north of 3rd Armored Division, the US Army's 1st Armored Division ran into the northern elements of the Tawakalna Division on the afternoon of 26 February. The fighting continued through the evening, with 21 Iraqi tanks and 22 APCs knocked out by Abrams and Bradley fire. The fighting was often at very close quarters; four M1A1 Abrams tanks were disabled by tank and RPG fire. While the 1st Armored was dealing with the remnants of Tawakalna, the RFGC Medinah Division was moving forward to establish defensive positions along a ridge line west of the Rumalyah oilfields. The 1st Armored decided to attack with all three of its brigades on line. The 2nd Iron Brigade had all three of its M1A1 tank battalions moving forward on-line when it encountered the Medinah Division's 2nd Brigade strung out for 9 kilometres behind the 'Medinah Ridge' around 1300 hrs. The Abrams battalions began engaging at ranges in excess of 3000 metres, beyond the effective range of the T-72. 'The first seven to ten minutes were like no Grafenwoehr[1] I've ever seen because each company as it came over the horizon began engaging targets to their front. And I could not visualize the length of a target that would allow that many tanks to shoot!' Within 40 minutes, 60 T-72s and nine T-55s had

[1] Grafenwoehr in the main live-fire gunnery range for US tankers in Germany.

been hit along with uncounted BMPs and other vehicles.[1] 'All along the ridge, you saw things that looked like blow-torches', recalled one of the tankers, referring to the horrific ammunition fires in the Iraqi T-72s. In about one hour of fighting along the ridge, the division's Abrams tank battalions had destroyed 137 armoured vehicles. By the end of the day, the total was 186 Iraqi tanks and 127 APCs. The US Army's VII Corps was credited with destroying 1350 tanks and 1224 armoured infantry vehicles. Losses amounted to nine M1A1 tanks destroyed and four damaged. Fratricide and mines caused more casualties than Iraqi tanks.

Although the heaviest tank fighting took place to the north-west of Kuwait, in Kuwait itself, there were a number of tank-vs-tank engagements in the outskirts of Kuwait City International Airport. Tiger Brigade from 2nd Armored Division helped spearhead the Marine drive into Kuwait and is perhaps best known for sealing off the highway leading from Al Jahra to Iraq, the infamous

*The BCS also includes a mine roller, patterned after Soviet systems. This photo was taken during tests at Aberdeen Proving Grounds after the right roller had detonated an anti-tank mine. (US Army)*

'Highway of Death'. Iraqi tank forces in Kuwait were equipped mainly with Chinese Type 59 and Type 69 tanks which proved to be completely inadequate against the newer M1A1 tanks. One of the few encounters between the newer T-72 tanks and M1A1s inside Kuwait involved Bravo Company

[1] The presence of T-55s with the T-72s was due to the unusual Iraqi practice of using T-55 command tanks in T-62 and T-72 tank regiments.

*A side view of the M1A2 Abrams. This version is most easily distinguished by the CITV system, mounted immediately in front of the loader's station. (GDLS)*

*A frontal view of the M1A2 Abrams during trials. In this view, two of the key external changes: the ICWS cupola and CITV sensor can be seen. (GDLS)*

of the 2nd Marine Tank Battalion, a Marine Reserve unit which had converted to the M1A1 shortly before the outbreak of the ground war. On the night of 24-25 February, the company was 'coiled up' in night defensive positions to give the crews some sleep. Around 0550 hrs on 25 February, a gunner on watch spotted movement through his thermal sight. It was a T-72 tank battalion of the 3rd Saladin Armored Division. An Iraqi lieutenant who survived the engagement recalled: 'Our column was headed across the desert when all of a sudden, the tank in front of me, to the left of me, and behind me, all blew up.' The lieutenant ordered his crew to abandon their T-72M. But it was hit before they could act, hurling him into the air and killing the other two crewmen. The firefight lasted 90 seconds by which time 34 out of 35 Iraqi tanks had been destroyed. In the following minutes, surviving Iraqi BTR-63s[1] and other vehicles tried to flee, only to be hit at long range. During the engagement, one M1A1 crew was credited with seven T-72s for seven rounds of ammunition.

*A close-up detail shot of the ICWS Improved Commander's Weapon Station which uses new wide-angle, laser-hardened periscopes as well as a simpler .50 cal machine gun mounting*

[1] BTR-63 is the Iraqi name for Chinese YW531 (Type 63) armoured personnel carriers.

## The Battle of Rumalyah

The last major tank-vs-tank engagement of the ground campaign occurred several days after the cease-fire. The US Army's 24th Infantry Division was serving as the 'cork-in-the-bottle', sealing the remaining Republican Guard units from escaping north into central Iraq. In the early morning of 2 March 1991, elements of the RFGC 1st Hammurabi Division and other units in the Basrah pocket attempted to break out in spite of the cease-fire. They began firing on Bradleys from Task Force 2-7 Infantry. In response, TF 2-7 Infantry was given permission to return fire. The causeway over the Hawr al Hammar waterway was sealed off by artillery and MLRS strikes. The 1-24th Aviation 'Viper' Battalion sent its 18 Apache helicopters into the area. The Apaches destroyed 32 T-72s, 49 BMPs, two ZSU-23-4 Shilkas and 48 other vehicles. In the meantime, the M1A1 Abrams tanks of 4-64 Armor passed through TF 2-7 Infantry and began to attack the remainder of the Hammurabi Division in the Rumalyah oilfields. By late afternoon, the Iraqis had lost 187 armoured vehicles, 34 artillery pieces, 400 wheeled vehicles and seven FROG missile launchers. One M1A1 Abrams was lost after it was set on fire by a T-72 exploding alongside it.

## The Lessons of Desert Storm

The lop-sided nature of the Allied victory in the Gulf War has led many army leaders to be extremely cautious about using Desert Storm to predict the nature of future conflicts. Nevertheless, at a technical level, the war has a number of lessons. The most obvious lesson is that technology alone does not determine the outcome of battles. The Iraqi's were poorly trained, poorly led and weakened by incessant air strikes. The Allied tank units were highly motivated, well trained and well led. Older US Marine M60A1 tanks performed extremely well against Iraqi tanks, including the newer T-72s. Still, technology advances in the M1A1 Abrams helped to shatter Iraqi tank units and minimise US losses. The Iraqi T-72 proved to be distinctly inferior to the M1A1 in all key areas: mobility, firepower and protection.

The one major innovation in tank combat demonstrated for the first time in Desert Storm was the thermal gunner's sight.[1] After the Korean War, the US Army had determined that the key tactical advantage in tank combat was to spot the enemy first and to engage him first. Since Korea, the US Army had placed great stress on target acquisition technology and has pioneered all the significant advances in tank sights; first the image intensification night sight in the 1960s, then the thermal imaging sight in the 1970s, and today, the millimetre wave multi-sensor. Target acquisition had been the second priority in the M1 Abrams tank design, after crew survivability. Desert Storm proved that this approach had been correct.

*An IPM1 of the 24th Infantry Division during Desert Shield. This division was the first heavy division into Saudi Arabia and was still equipped with the older M1 and IPM1 tanks. This particular tank is fitted with the external APU at the rear. (US Army)*

Although designed originally as a night sight, the thermal sight has important advantages in day combat also, especially in conditions of poor visibility. Much of the ground fighting during Desert Storm took place in dismal weather conditions: low lying clouds, sand storms, smoke from the Kuwaiti oil fires, battlefield fires and dust kicked up by vehicles and gunfire. Thermal sights allowed M1A1 tank crews to see through much of this murk. In contrast, the Iraqi tankers had no thermal sights and were blind in

---

[1] The M60A3 tank was the first tank in the world to be fitted with thermal sights, but the M1 Abrams was the first tank to be designed with a thermal sight in its fire control system from the outset.

*An M1A1 Abrams being off-loaded during the roll-over programme before Desert Storm. This vehicle shows several interesting details including the external APU and the additional access doors on the engine deck roof, characteristic of the M1A1. (US Army)*

many engagements. The first they knew of the presence of Abrams tanks was when one of their own vehicles exploded under the impact of a 120 mm projectile. The thermal sights allowed the Abrams to engage at ranges of over 3500 metres, beyond the effective combat range of the Iraqi T-72Ms.

This range advantage was not due to the thermal sights alone. Although the ballistics of the US 120 mm gun and Iraqi 125 mm gun are not substantially different, the M1A1 Abrams tanks were consistently able to engage the T-72s at ranges far in excess of the effective range of the Iraqi tanks due to training and fire control advantages, not raw gun performance. US tanks typically began engaging Iraqi tanks at ranges of 3000 metres or more if visibility permitted, and enjoyed high hit rates even at these ranges. US Army tankers have repeatedly spoken about Iraqi tank fire falling short. The poor performance of Iraqi tankers was due not only to poor training but inherent limitations in Soviet tank gun design philosophy. To minimise costs, Soviet tank guns (and tanks in general) are designed with low life expectancies. Barrel life on the 125 mm D-81TM gun is 120 rounds before replacement is needed. On the US M256 120 mm gun, it is over 1000 rounds. To minimise peacetime operating costs, peacetime gunnery training is very limited on Soviet tanks, typically less than a dozen rounds per year per crew versus over a 100 on US tanks per year per crew. Technology affects training which then affects combat performance. Iraqi tanker training, based around Soviet tank technology and training practices, was appallingly poor. Although M1A1 Abrams destroyed about 500 T-72s by gunfire, there are only seven known cases of T-72s hitting M1A1 Abrams tanks.

The M1A1 and M1A1HA armour worked extremely well as did the internal ammunition compartmentalisation. At least seven M1A1 Abrams took direct hits by 125 mm projectiles. None penetrated and no tanks were disabled. In one instance, an M1A1 was hit by two 125 mm APFSDS projectiles fired in rapid succession from only 500 metres away. One hit the hull front, one the turret front and there was no injury to the crew or debilitating injury to the vehicle. A nearby tank was hit about the same time on the front of the turret side, located the T-72 and destroyed it at close range before the T-72's autoloader could get off another shot. Official sources indicate that there were 18 cases of combat damage to M1A1s. In all nine cases of permanent losses, the cause was fratricide from other US weapons. The other nine cases of damage was primarily due to mines and all vehicles were considered repairable. Reportedly, two M1A1s were set on fire and destroyed by their crews when they had to be abandoned. No M1 tankers were killed within the protective armour of the tank by enemy fire.

*A tank attack by a VII Corps unit during Operation Desert Storm. Much of the ground fighting took place under poor weather conditions or at night, which worked to the advantage of US tankers. (US Army)*

*The CITV Commander's Independent Thermal Viewer on the M1A2 gives the commander his own day-night sensor to acquire targets (GDLS)*

Casualties amongst M1A1 crews did occur, such as a tank commander hit by artillery fragments while riding outside the hatch. On the several occasions where the ammunition bustles in the M1A1 were hit, the compartmentalisation prevented the crew from suffering the consequences of the ammunition fires.

In complete contrast, the Iraqi T-72 tanks proved horribly vulnerable to catastrophic ammunition fires. The armour of the T-72 was inadequate to protect the tank from 120 mm fire, and some Abrams crews claimed that their 120 mm projectiles went through sand berms, through the T-72's frontal armour, through the engine compartment, and out of the rear. In most cases, penetration of the T-72 by either sabot or HEAT projectiles led to catastrophic ammunition fires. Frequently, sabot rounds would enter the turret or hull and shatter one or more of the 125 mm propellant cases. Because of the pyrophoric effects of depleted uranium penetrators when they pass through steel, this would almost inevitably lead to the propellant being ignited. Once the propellant began to burn, the fire often quickly spread to neighbouring ammunition. If the chain reaction was quick enough, the resulting explosion would blow the turret off the T-72. Even if it did not lead to a rapid chain reaction, the ammunition fire would inevitably cause a massive internal fire, which American tankers described as 'blow-torches' or 'furnaces'. Abrams tankers often remarked that the T-72s seemed to lose their turrets more often than older T-55s or T-62s, probably because of the larger volume of ammunition propellant stowed in their small hull interiors.

*An M1A1 named 'Final Option' of F Troop, 3 ACR during Operation Desert Storm. This regiment used troop letters rather than chevrons for unit identification. (US Army)*

The lethality of existing tank guns was not fully appreciated until Desert Storm. Although the theoretical ballistic parameters of these weapons were well understood, the devastating results of these weapons in combat surprised many of their crews. The new M829A1 120 mm APFSDS round was popularly dubbed the 'Silver Bullet' by M1A1 Abrams crews for its outstanding performance.

The overall performance of the M1A1 Abrams was extremely satisfactory in the eyes of its crews. The M1 Abrams had been unfairly ridiculed as a 'lemon' by the US mass media in the early 1980s, but this was not a view widely shared by the Abrams crews before the war. Reliability was extremely high, with readiness rates exceeding 90 per cent. For example, during the long march to the Euphrates, 4-64 Armor suffered one breakdown, a faulty fuel pump, which was repaired in 45 minutes. Media stories about the vulnerability of the Abrams to sand ingestion proved inaccurate. Worries about the Abrams high fuel consumption were allayed by careful planning and the use of the new HEMTT fuel tankers. Capt. David Hubner, a company commander with 1-64 Armor, described the usual drill:

'We would go approximately 100 to 150 km depending on the terrain and we would stop for fuel or maintenance. The crew gets cramped up after two or three hours so you want to get out and stretch. At that time we'd check the track. We'd fuel up. We'd blow out the air filters. We've got wands that attach to the back of the engine to blow out the filters and we'd do that religiously to keep the tank running. I took 13 tanks across the border, and brought 13 back without any problems.'

The Desert Storm tank battles emphasised the effectiveness of combined arms tactics. In describing Desert Storm battles, it is no longer to possible to speak about 'tank' battles. Unlike armour

*A US Marine Corps M1A1, probably from 2nd Tank Battalion passes by a revetted Iraqi vehicle during Operation Desert Storm. The 2nd Tank Bn. was the only Marine tank unit fully equippped with the M1A1 during the fighting. (US Marine Corps)*

*An M1A1 with typical VII Corps markings during Desert Storm. (Joel Paskauskas II)*

engagements in all previous wars, Bradley infantry and cavalry fighting vehicles were almost always present in significant numbers and took part in the exchanges with enemy armour. Likewise, it was rare to see a homogeneous Iraqi tank unit without BMPs or other armoured vehicles. Heavy divisions now integrate all the combat arms – tanks, infantry, scouts, engineers, and artillery – under armour. With the advent of new artillery projectiles such as the DPICMs[1], artillery played a frequent role in the tank fighting. Attack helicopters are starting to show signs of revolutionising land warfare. The weather on several occasions prevented Apaches from intervening in several of the large tank battles. But when Apaches did intervene, such as at the Rumalyah oilfields on 2 March, they decimated the Iraqi armour and fatally weakened the Iraqi formations before the Abrams tanks intervened.

The problem of fratricide, or 'friendly fire' was underestimated before Desert Storm. A total of 81 soldiers and 77 per cent of US Army matériél losses (including seven M1A1 Abrams tanks and 20 Bradleys) were due to friendly fire incidents. The terrain conditions in Iraq, combined with advances in fire control, permitted engagements at ranges far in excess of those possible in previous wars. At 3000 metres, a tank simply looks like a blob of light on a thermal imaging sight whether it's a T-72 or M1A1. Fratricide casualties seemed a greater problem in Desert Storm than in previous conflicts if only because casualties from enemy action were so incredibly small. Steps are now being taken to improve command and control to identify more accurately where friendly forces are located. Identification-Friend or Foe systems are notoriously difficult to design since they not only distinguish friends from enemies, but they help identify friends to the enemy.

In the wake of Operation Desert Storm, main battle tank development is likely to slow down. The outstanding technical performance of the M1A1 tanks and the dismal performance of the Soviet T-72M reduces the incentive to develop more advanced designs. It is important to note that the T-72s encountered in Operation Desert Storm were not the best tanks built by the Soviet Union. The best Soviet tanks such as the T-72B1 and T-80U have not been exported. These tanks have significantly better armour protection, and modestly superior fire control systems to the T-72Ms encountered in the Gulf. But the collapse of the Soviet Union, and the slow-down of its tank industry, is likely to remove one of the other competitive incentives in the development of tank technology. The M1A1 and improved derivatives like the M1A2 Abrams are likely to represent the peak of American tank technology until the end of this decade.

[1] Dual-Purpose Improved Conventional Munition. These rounds contain several dozen small shaped-charge submunitions which can penetrate the thin roof armour of armoured vehicles.

# M1 ABRAMS VARIANTS

The M1 Abrams has not been extensively used to form the basis for special purpose vehicles. The hull of the vehicle is relatively expensive and there has been the feeling that special applications such as armoured recovery vehicles or combat engineering vehicles do not need a chassis with the high speed capabilities of the M1 tank.

General Dynamics developed an armoured recovery vehicle on the basis of the M1 to satisfy a US Army requirement for a vehicle to replace the M88A1 recovery vehicle in M1 tank units. The US Army has generally favoured the idea of a modernised M88 version, the M88A2, due to its lower cost. However, there were complaints about the slow road speed of the M88 during Desert Storm which may lead to the issue being re-opened at a later date.

The M1 chassis has also been used as the basis for a Heavy Assault Bridge carrier. This system is designed to supplement the AVLB currently based on the M60 chassis. None of these have been funded to date.

Tanks are frequently used to carry forward observers and forward air controllers (FO/FAC) into battle to help co-ordinate air support for tank units. A programme has been under way since 1991 to develop changes in the M1A1 Abrams to allow FO/FAC officers to deploy in the tanks with the necessary communications equipment. The FO/FAC version will be identical to normal tanks, but will have the brackets and hardware necessary for the observers' communication equipment.

# THE PLATES

M1 tanks were originally delivered from the factories in overall FS 34079 Forest Green. Any camouflage painting was applied subsequently in the unit depots. Originally, this involved the application of one of the optional four-colour MERDC patterns. These are detailed in Osprey Vanguard 41. In 1987, the new NATO three-colour scheme began to be applied at the factory. This consists of FS 30051 Green, FS 34094 Brown and FS 37030 Black. This new finish was applied with a special type of paint that could be cleaned with chemical

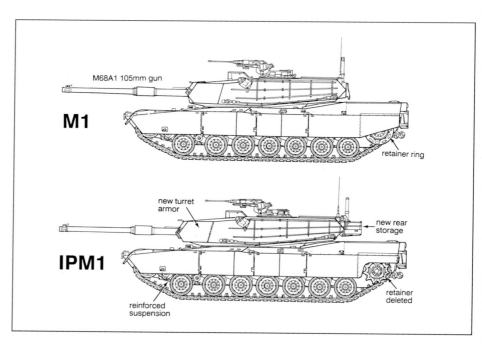

*These diagrams show the main differences between the M1 and the IPM1 (Improved M1) variant. October 1984 saw completion of the first IPM1 with final deliveries made in May 1986. Production switched solely to the M1A1 after 894 Improved M1s had been built.*

agent decontamination solutions without peeling. Many units repainted their older M1 tanks in the new NATO scheme during depot overhauls. The pattern of this scheme is shown in Osprey Elite 26.

During Desert Storm, US M1A1 tanks were finished in FS 33446 Tan, also known as CARC Tan 686. Like the paints for the new NATO scheme, this is a special permanent paint which does not dissolve when subjected to cleaning by common chemical warfare decontamination solutions. The US Marine Corps uses a paint of the same colour, Desert Tan 686, which does not have the same resistant qualities. However, the majority of Marine M1A1 tanks came out of army inventories with the army paint.

Basic markings during Operation Desert Storm followed usual US Army practices. Prior to the ground campaign, most vehicles had 'bumper codes' on the front and rear which give basic unit data on the left and company/vehicle number on the right. The numbering pattern for line combat vehicles is generally: 66 (battalion commander), 65 (battalion executive officer); 11, 12, etc. (1st Platoon), 21, 22, etc. (2nd Platoon), 31, 32, etc. (3rd Platoon). These bumper codes were overpainted before the fighting began in some units. Most units adopted some form of large unit marking to assist in identifying tanks in the field. These generally were painted in black on the front of the side skirts and on a detachable panel on the rear of the turret. The US Marines had adopted a simple system of chevrons for company identification before the war, probably inspired by similar Israeli

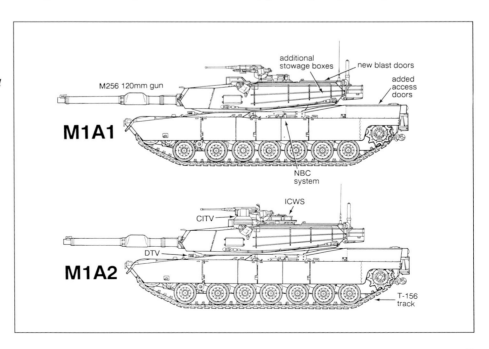

*Top right: The Abrams Leguan HABS features a horizontal launch, 26 m MLC 70 (Military Load Class) bridge on a converted M1 MBT chassis*

*Right: The M1 continues to evolve, these diagrams showing the latest developments in the Abrams family. The main improvements are the larger M256 1200 mm gun on the M1A1 and in addition on the M1A2 the ICWS, CITV and DTV.*

*An M1A1 named 'AMEN' from VII Corps during Desert Storm. (Joel Paskauskas II)*

*An M1A1 of 2nd Platoon, Charlie Company, 4-64 Armor, 24th Infantry Division (M) during the Battle of the Rumalyah Oilfields on 2 March 1991. This was the last major tank battle of the war, when Iraqi Republican Guard mechanised units tried to push out of the Basra pocket after the cease-fire. (24th Infantry Div. PAO)*

markings, and sometimes called the 'Spinning Vee'. The chevrons rotate sequentially in a clock-wise direction: A Company (up); B Company (facing right); C Company (down); D Company (facing left). In some cases, all four Vees were combined in a diamond shape, probably indicating headquarters tanks. This system was adopted by most tank units of VII Corps. However, in some units the order of the chevron seems to have been deliberately changed, perhaps as a security measure. In the US Army VII Corps, the chevron could have a two digit number added inside the chevron. This appears to have been a battalion code-number (first digit) followed by a platoon number. Some exceptions are detailed below. Often, only the single battalion number was painted on the side, but both numbers were carried on the rear turret identification plate. The Marines used a small square inside the chevron to identify the platoon. The units of XVIII Airborne Corps had their own markings which are described in the plate notes below.

All tanks had a large black upward chevron

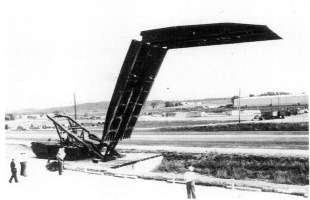

*General Dynamics developed an Abrams Recovery Vehicle with company funding based on the M1 chassis due to acknowledged shortcomings in the existing M88 vehicles when towing the M1. The programme to replace the M88 has been controversial, with the army preferring modified versions of the M88 due to their lower cost. (GDLS)*

*The first version of the Heavy Assault Bridge on the M1 chassis, developed by BMY, entered tests in 1984 as is seen here. The army later decided a different type of bridge was needed and a General Dynamics alternative with the MAN Leguan bridge was selected. (BMY Corp.)*

painted on the side-skirts shortly before G-Day, the accepted recognition marking for the Allied forces. In addition, a VS-17 identification panel was tied to the rear roof or engine deck of the tank, often over the turret stowage. This is a 70 x 30in. plastic sheet, fluorescent orange/red on one side and white on the reverse with green canvas reinforcement and tie-downs along the edges. It has been the standard method of aircraft recognition in the US Army since World War 2, and alternate colours include fluorescent yellow and blue.

**Plate A:** *M1A1 Abrams, A Co., 1-37 Armor, 1st Armored Division, KTO, January 1991.*

Appropriately enough to start the plates, the 37th Armor traces its lineage back to the 37th Tank Battalion, commanded by Creighton Abrams in World War 2. The scheme is typical for a VII Corps tank with the chevron marking. Several special markings were adopted in the battalion, notably a bow marking showing a map of Iraq penetrated by a sabot projectile. The bumper codes are 1^1-37 A-31. Unit tanks had cartoons on the bore evacuator, a shark in this case, and a simplified version of Frank Franzetta's famous illustration Death Dealer in the case of B Company. The Franzetta painting was very popular amongst tankers stationed in Germany, and was a motif used frequently in tank insignia during Desert Storm.

**Plate B:** *M1A1 Abrams, 3-66th Armor, 1st Infantry Division, KTO, January 1991.*

1st Infantry Division (M) was the exception to the rule so far as the Spinning Vee insignia was concerned. In their tank battalions, the chevron indicated platoon rather than company: 1st platoon (upward), 2nd (right), 3rd (down), 4th (left). The two digit number indicated battalion and company.

*One of the odder M1 variants was the SRV surrogate research vehicle. This vehicle tested new tank vision devices in anticipation of later programmes when unconventional turret configurations would be developed. (US Army)*

*The second bridging version of the M1 Abrams developed by General Dynamics uses the MAN Leguan heavy assault bridge. (GDLS)*

The 1st and 2nd Brigade were sequential: 1x (5-16 Infantry); 2x (1-34 Armor); 3x (2-34 Armor); 4x (2-16 Infantry); 5x (3-37 Armor); 6x (4-37 Armor). The division's third round-out brigade came from 2nd Armored Division, which restarted the numbering pattern: 1x (1-41 Infantry); 2x (2-66 Armor); 3x (3-66 Armor). The second digit of the pair indicated company: x1 to x6 (A to E Company); x6 was HHC Company. So this vehicle is 2nd Platoon; B Company; 3-66 Armor.

**Plate C:** *M1A1 Abrams, 3-8 Cavalry, 3rd Armored Division, KTO, February 1991.*
The 2nd Brigade of 3rd Armored Division used one variation from the usual VII Corps pattern: the skirt markings were painted in sand on a black square. Many of the unit's tanks had a small marking painted on the skirt on a black rectangle, this is believed to have been a version of the division's 'Spearhead' insignia.

**Plate D:** *M1A1HA Abrams, H Troop, 3rd Armored Cavalry Regiment, Operation Desert Storm, February 1991.*
This cutaway drawing shows the basic internal layout of the M1A1 Abrams. The internal layout of M1, IPM1 and M1A1 are all essentially similar although continual improvements were incorporated through the production run. The driver's station in the front of the tank is self-contained, with the driver lying in an almost prone position when the hatch is closed. This posture was adopted to minimize the height of the hull, and hence, the overall vehicle height. Control is via a yoke control evident in the illustration. On either side of the driver are fuel cells. The turret houses the crew, armament and fire controls. This illustration does not attempt to depict the layout of the Burlington special armor package as this is still a compartmented top secret subject. This armor is quite thick; even though the M1A1 turret is noticeably larger on the outside than the older M60A1 tank, internally, the M1A1 is smaller due to the space taken by the armor package. The tank commander sits under the main right side hatch, with a remotely operated M2 .50 cal machine gun in front of his hatch. In front of him is the gunner with his large array of gun sighting devices and fire controls. The thermal imaging sight, optical day sight and laser rangefinder are contained in an integrated package in the armored box in front of the commander's cupola. The loader is stationed on the left side of the breech. In front of him is a

variety of stowage including ammunition for the co-axial machine gun, and radio stowage. The main tank gun ammunition stowage is located in the rear bustle of the turret behind special blast doors. The loader's hatch is fitted with a pintle mounted M240 7.62 mm machine gun no illustrated on this plate. The loader is responsible for air watch and overwatch during vehicle travel. The main vehicle powerpack, the AGT-1500 1500 horsepower gas turbine engine, is located in the hull rear. This is not shown in the plate except for one of the rear sponson fuel cells. The engine and transmission are an integrated design to permit easier replacement in the field.

**Plate E:** *M1A1 Abrams, B Squadron, I Troop, 3-2nd Armored Cavalry Regiment, Battle at 73 Easting, February 1991.*
The wartime markings had the upward chevron identification marking added. The black stripes on the barrel are probably platoon markings. The green stripe on the turret front is a strip of velcro tape used to attach the MILES laser simulation equipment during training.

**Plate F1:** *M1A1, D Troop, 1-3 Armored Cavalry Regiment, KTO, February 1991.*
The 3rd Armored Cavalry Regiment was one of the few units not to use chevrons, and instead painted large troop numbers on the side skirts and panels, usually followed by a company number. This battalion also used a Hell's Dragon insignia on the turret front. Evident on the rear roof is the usual orange ID panel.

**Plate F2:** *M1A1, I Troop, 3 Armored Cavalry Regiment, KTO, January 1991*
Road Beast, a mine rake vehicle with I Troop, had cartoonish eyeballs painted on the plough blades, as well as a popular taunt painted on the barrel. Iraqi mines proved little of an obstacle to vehicles such as Road Beast.

**Plate G1:** *M1A1, C Company, 4-64 Armor, 24th Infantry Division (M), Battle of the Rumalyah Oilfield, 2 March 1991.*
24th Infantry Division (M) had its own markings system from before the war which differed from the VII Corps system. In this case the chevrons were chosen to resemble the company letter. So A Company used an upward pointing chevron, a rightward pointing chevron denoted B Company, a leftward pointing chevron C Company and a triangle was D Company. As in the Marine case, the 24th Infantry Division used small squares to identify platoons. Many tanks had another downward pointing chevron on the turret front. This symbolised V for victory, the 24th Infantry traditionally being called the Victory Division.

**Plate G2:** *M1A1, B Company, 2nd Marine Tank Battalion, KTO, February 1991.*
The inset drawings showing the usual Marine chevron pattern. For unknown reasons, this company did not use the markings in the standard fashion. Insignia include a palm tree with Marine Corps globe and anchor, inspired by the Afrika Korps insignia, and kill markings on the turret front and thermal sight housing.

# INDEX

(References to illustrations are shown in **bold**. Plates are shown with caption locators in brackets.)

1st Armored Division 19, 34–35
1st Cavalry Division **10**, 17, 19
1st Infantry Division 20, 22, **A** (45), **B** (45–46)
2nd Armored Cavalry Regiment **E** (47)
2nd Armored Division **7**, 18, 23–24, 35, 46
3 Armored Cavalry Regiment 20, **40**, **D** (46–47), **F** (47)
3rd Armored Division 19, 33–34, **C** (46)
3rd Infantry Division **9**
7th Corps 19–20, 21, 23, 35, **38**, **41**, **44**, 44
11 Armored Cavalry Regiment **21**, **22**
18th Airborne Corps 20, 21, 44
24th Infantry Division 17, 20, **23**, **37**, 37, **44**, **G1** (47)
73 Easting, battle of 23–24, 33, **E** (47)

Abrams, Creighton 8
airstrikes, Operation Desert Storm 21
ammunition 3, 4, 9–10, **16**, 39, 41, **D** (47)
anti-tank weapons 5–6, 16, 18
armament 3–4, 6, 7, 9, 10, 14, 34, 38, 40
armour 5, 6, 9–10, 11–12, 38, **D** (46)
auxiliary power unit (APU) 11, **38**

Battalion Countermine Sets (BCS) 21, **23**, **33**
Battlefield Combat Vehicle Identification (BCVI) system 16
bumper codes 43

camouflage **11**, **12**, **19**, **22**
casualties 38–39, 41
Challenger, British tank 3, 12
Chrysler 4, **5**, **6**, 6, 8
colour schemes 42–43
combat actions, Operation Desert Storm **21** (map)
Commander's Independent Thermal Viewer (CITV) 10, 13, 15, **34**, **35**, **36**, **39**
commander's independent weapons station (ICWS) 13, **36**
Components Advanced Technology Test-Bed (CATTB) 14
cost 3, 4, 7, 10

depleted uranium penetrator rounds 9, 39
deployment, Saudi Arabia 16
design requirements 3

Egyptian forces 5, 6, 12
electronics system 13
engines 4, 7–8, **18**
enters service 8

fire control 10
FMC Corporation 7

fratricide 38, 41
fuel consumption 11, 40
funding 13, 14, 15
future developments 13–14

General Dynamics Land Systems Division (GDLS) 8, 42
General Motors 4, 6, 7–8
Germany 3, 6, 7
grenade launchers 15–16
Gulf War, the, 1987–88 17

Heavy Assault Bridge Carrier 42, **43**, **45**, **46**
helicopters 41
'Highway of Death' 35

identification panels 45
improvements 15–16
 Block I: 11–12
 Block II: 13
 Operation Desert Storm 16–17
integrated command and control system 13
interior **13**, **14**, **15**, **24**, **34**, **D** (46–47)
IPM1 10, **17**, **18**, **19**, **37**, **42**
Iraqi forces 16, 17, 18, 20–21, 22, 23, 37, 37–38
 at 73 Easting 23, 24, 33
 losses 16, 22, 23, 34, 35, 36, 37
 at Medinah Ridge 34–35
 at Wadi al Batin 33–34
Israeli forces 5, 6

Kuwait 13, 35, 35–36
Kuwait Theatre of Operations (KTO) 18, 19–20

Laser Warning Receiver (LWR) 16
Leopard II, German tank 3, 7, 12
lessons, Operation Desert Storm 37–41
losses 35

M1A2: 12, 13, 15, **35**, **36**, **43**
M1E1: 10, 11
M60 series main battle tanks 3, 5, 8
Marine requirements 12, **33**
markings 43–45, **A** (45), **B** (45–46), **F1** (47), **G** (47)
MBT-70 3
MBT task force, the 3
media denunciation 8, 40
Medinah Ridge 34–35
minefields, Operation Desert Storm 21–22
mines and mineclearing 21, **23**, **33**, **35**, 38, **F2** (47)

NATO 3, 6–7
navigation systems 13, 17

nuclear-biological-chemical (NBC) protection system 10

official designation 8
organisation 18, 18–19

Pakistan 12
performance, Operation Desert Storm 16, 34, 35, 40
POMCUS (Prepositioned Matériél in Unit Sets) 11
procurement 14 (table)
production 8, 11, 12, 13
prototypes **4**, **4**, **5**, **6**, 6

recovery vehicles **17**, **18**, 42, **45**
Reforger exercises **9**, **11**, **12**, **22**
reliability 40
Rumalyah, battle of 37, 41, **44**, **G1** (47)

Saudi Arabia 12, 13, 16, 16–17, 17–18
scale drawings **42**, **43**
skirts **7**
Sleep Support System (SSS) 17
Soviet forces 6, 9, 38
specifications: **D**
stowage racks **20**, **21**
Surrogate Research Vehicle, the 14, **45**
survivability 4, 8
Syrian forces 5

T-62, Russian tank 5
T-72, Russian tank 23, 24, 37, 38, 39, 41
tactics, Operation Desert Storm 40–41
tank design, US Army 3
technological developments 3
thermal gunner's sight 37–38
tracks 17
training **10**, 17, 38

US Army in Europe (USAREUR) **9**, 11
US Marine Corps 12, 18, 19, 20, 21, **33**, 35–36, 37, **40**, 43, 44–45, **G2** (47)

variants 42, **43**, **45**, **46**

Wadi al Batin 33–34
wading trunks 12

XM1 **4**, **5**, 5–8, **6**, **7**, 10
XM1E2 Close-Combat Test-Bed (CCTB) 14
XM803: **3**, 3
XM815: 3–4

Yom Kippur War, 1973: 5, 5–6